12/25/77

To Richie,
Merry Christmas!
All my love.
Peggy

FIRE ENGINES AND FIRE-FIGHTING

FIRE ENGINES AND FIRE-FIGHTING

David Burgess-Wise

LONGMEADOW

CONTENTS

First published in the USA 1977 by
Longmeadow Press, P O Box 16
Rowayton Station, Norwalk
Connecticut 06853

ISBN 0 7064 0613 3

Produced by Mandarin Publishers Limited
22A Westland Road
Quarry Bay, Hong Kong

Printed in Hong Kong

FROM THE BURNING OF ROME TO THE GREAT FIRE OF LONDON

As soon as primitive man realized that fire was not a visitation from the gods, but a force that could be harnessed for his own benefit, the search began for some reliable means of controlling fire's destructive capability.

The first recorded type of firefighting apparatus appears to have been nothing more than a leather bag with a long nozzle attached, which was filled with water and squirted at the blaze. Obviously, both its range and power were extremely limited, and it could be used to combat only the most minor domestic outbreaks.

Around 300 BC the Greek scientist and philosopher Ctesibus developed a double-cylinder force pump – he was working in the Egyptian city of Alexandria, scientific think-tank of the ancient world, and had previously conceived inventions as diverse as a water-clock and an hydraulic organ. This pump must have been an impressive device, for two other ancient scientists, Hero of Athens (who first discovered the power of steam) and Vitruvius, both mentioned it at some length in their writings.

Hero mentioned that Ctesibus's pump could be used for firefighting, as its delivery pipe was so jointed that it could be directed both up and down and from side to side. Vitruvius described the pump as having cylinders and valves of brass, while the pistons were packed with unshorn sheepskin to make them watertight.

Between the cylinders was 'a basin having a cover like a funnel . . . which is adjoined and fastened . . . by a collar riveted through and on the top of it a pipe is affixed vertically'.

Pipes led from the cylinders to an intermediate chamber, while, at 'the upper holes of the pipes within the basin are made valves, hinged with very exact joints which, stopping the holes, prevent the efflux of the water that would be pressed into the basin by air . . . the cylinders have valves placed below the lower mouths of the pipes and fixed over holes that are in their bottoms'.

The Great Fire of London 1666 – more than 13,000 houses were destroyed over an area of 436 acres

Two men were needed to operate the pump, alternately pushing down on levers attached to the pistons. Each down-stroke 'pressed the air therein contained with the water through the mouths of the pipes in the basin, from whence rising to the cover, the air pressed it upwards through the pipes'.

When this description, after being lost for over 1800 years, was republished in a German technical treatise of the sixteenth century, it was for a long time contended that the last sentence meant that an air pressure vessel was fitted to the pump so that the water could be discharged in a continuous stream: but it's likely that this was a misinterpretation of the original text, and that Ctesibus's pump laboured in irregular fits and squirts.

The ancient Romans had force-pumps, too, and one of these is preserved in the British Museum in London. Roman cities presented a tremendous fire hazard, especially in the densely populated slum rookeries inhabited by the poor, where dilapidated buildings, constructed mainly of wood, huddled together. The situation was not helped by the Roman custom of keeping a fire burning day and night on the household altar to keep the domestic gods in a good humour.

Out of sheer necessity, the Romans created an organized fire service, the Corps of Vigiles, formed from selected male citizens who had to undergo firefighting training, and were then formed into companies, each one under the command of an officer called a *siphonarias*.

Armed with pumps, buckets, ladders and hooks for pulling blazing thatch from buildings, the Vigiles were supposed to stop fire from spreading. Judging from the number of times that Rome was ravaged by fire, the Vigiles were conspicuously unsuccessful in fulfilling this objective. In AD 64, for instance, the Eternal City burned for three days and nights, with great loss of property.

This was during the reign of the notorious Emperor Nero, who, far from 'fiddling while

Technological revolution 1673 – the first use of Van der Heiden's improved fire engines, at the Amsterdam Ropewalk

H
E
H
E
H
O
B
F

Rome burned', actually attempted to improve the city's firefighting forces and provide some measure of relief to the homeless. It was to whitewash the shortcomings in his administration that he laid the blame for the fire on the Christians.

Obviously, lessons had been learned from Nero's inefficiency, for, some 40 years later, the Emperor Trajan commanded the historian Pliny, who was at that time Governor of Bithynia, 'to provide such machines as are of service in extinguishing fires, enjoining the owners of houses to assist in preventing the mischief from spreading and, if it should be necessary, to call in the aid of the populace'.

Good intentions were, apparently, not enough: Pliny subsequently wrote to the Emperor to tell him that the town of Nicomedia in his province had been virtually destroyed by fire, 'owing to the violence of the wind and to the indolence of the people', who, it seems, stood fixed and idle spectators of this terrible calamity.

'The truth is', confessed Pliny (the younger, not to be confused with his uncle, who died at Pompeii), 'that the city was not furnished with either engines, buckets, or any single instrument proper to extinguish fires, which I have, however, now ordered to be provided'.

Organized firefighting died with the Roman Empire, and was not to be revived for many centuries. In Saxon England, for instance, most people lived in small villages, and it was only when larger communities were built within fortified walls to repel the invading Danish pirates that the spread of fire again became a hazard: London was devastated by fire in AD 798 and again in 982, thanks to firefighting methods which were cruder than those of the Romans. Grappling hooks to drag down burning thatch and buckets in which water could be passed from hand to hand along a chain of men from pond or river to the fire were the sum total of their achievements.

The most effective means of fire protection at this time was the curfew, by which all householders were compelled to extinguish their fires at sunset. Under the Normans (whose expression '*couvre feu*' was the origin of the 'curfew'), this became an effective means of preventing insurrection, preventing Saxons from meeting and possibly plotting against the Norman conquerors.

Mechanical means of fire extinction did not appear again until the sixteenth century, when the Portuguese were recorded as using fire squirts, big metal syringes about 1 metre long, equipped with handles at either side, each of which was held by one man: a third operator held the handle attached to the piston rod. The nozzle of the squirt was placed in a bucket of water, the piston rod drawn back to fill it, and the water then discharged at the fire. A larger version of this device, mounted on a wheeled carriage, was built in Augsburg, Germany, by the goldsmith Anthony Blattner in 1518.

It was crude devices of this kind which were mostly used at the Great Fire of London in 1666, though it seems as though more powerful engines were already in existence. In 1634, one John Batt published a book entitled *A Treatise on Art and Nature*, in which he wrote of 'Divers squirts and petty engins to be drawn upon wheeles from place to place for to quench fier among buildings, the use of which hath been found very commodious and profitable in cities and great townes'.

The Great Fire pointed up all the fatal errors of complacency that the inhabitants of London had made. In the early seventeenth century King James I had complained that his capital was more dangerous than Rome in the time of Caesar Augustus: but no-one had heeded his advice to demolish the old wooden-framed houses and replace them with brick buildings.

Consequently, when Thomas Farynor's bakery in Pudding Lane, near London Bridge, caught fire early in the morning of Sunday, 1 September, 1666, the flames found little difficulty in establishing a hold on the other houses in the street. From there, sparks were carried from rooftop to rooftop until the flames were well beyond the control either of squirts or 'petty engins'. It was eventually only by creating gaps too wide for the flames to cross by blowing up buildings across their path that the city authorities managed to get the blaze under control. By the time that the fire was extinguished, three days later, it had laid waste 436 acres and destroyed some £10,730,500-worth of property, a colossal sum in those days. Among the buildings lost were 13,200 houses, 86 parish churches, six chapels and St Paul's Cathedral. Needless to say, the Londoners failed totally in rebuilding those 436 acres. Here was their chance to plan a safer city: instead, the old pattern of crowded streets was followed in the work of reconstruction. Sir Christopher Wren, who designed the magnificent new St Paul's Cathedral, drew up a plan for a new London: it was rejected out of hand.

The Common Council of the City of London concluded complacently that the fire had been caused by 'the hand of God upon us, a great wind, and the season so very dry'. Within a few months, however, it was grudgingly admitted that God had not been entirely to blame, for a new code of building practice was drawn up, which attempted to ensure that houses were not such firetraps as before. More importantly, the Common Council drew up new fire prevention regulations, which divided the City of London into four quarters, each of which had to be provided with 800 leather buckets, 50 ladders (from 12 to 42 feet long), 24 bucket sledges and 40 shovels. Moreover, the 12 main Livery Companies

(Opposite) Insurance company fireman and manual engine of 1805 – some of the crew stand inside the machine, assisting the pumpers by working treadles

(Right) The type of fire squirt used at the Great Fire of London

(Below) Firemarks such as this were affixed to houses insured against fire from the end of the 17th century

were ordered to provide an engine each, plus 30 buckets, two squirts and three ladders; lesser Companies had to provide what firefighting apparatus they could. Additionally, each parish had to have at least three brass squirts: everything was to be 'ready upon all occasions'.

Rules to ensure the early detection of fires were also published:

> These things being made and done, then the sentinel hath a place on the top of the highest steeple whereby he may look all over the town: and every two hours of the night he plays half an hour upon a flageolet being very delightful in the night: and he looks round the city; if he observes any smoke or fire or danger of fire, he presently sounds a trumpet and hangs out a bloody flag towards that quarter of the city where the fire is.

And if hanging out bloody flags should prove ineffective, citizens could now insure their property against fire damage: the Fire Office, founded in 1667, was the brain-child of Doctor Nicholas Barbon, whose sober name concealed the fact that he was the son of that colourful Cromwellian 'Praise-God Barebones', and had originally been Christened 'If-Jesus-hadst-not-died-for-thee-thou-hadst-been-damned!' The success of the

An early fire-pump, mounted on sleds and equipped with a gooseneck monitor

venture was immediate, for London was still far from being a fireproof city: in 1676 some 600 houses were burned down in Southwark, while six years later more than a thousand houses in Wapping were destroyed by fire. Other fire insurance offices were formed: but all found that they were paying out more in the way of premiums than was necessary, for once the initial enthusiasm had worn off, the ambitious plans for providing London with a comprehensive fire service ground to an apathetic halt.

So the fire insurance companies began to form their own firefighting companies, recruiting 'watermen and other lusty persons' as part-time firemen, who were paid for attending fire drills as well as putting out fires: perhaps almost as strong an inducement was the fact that a member of a fire brigade was automatically immune from the unwelcome attentions of the press gangs who roamed the streets forcibly 'recruiting' able-bodied men into the Royal Navy. Barbon wrote in 1684 that his Phenix Brigade were like 'Old Disciplined Souldiers, that do greater things then Ten times that Number of Raw and Unexperienced Men'.

The insurance companies began issuing policy-holders with metal 'firemarks' which were fixed to the wall of the house: and at first only the brigade of that company would fight fires at those premises. Gradually, however, brigades began to compete to attend fires, vying for the honour of having reached the blaze first.

Fire pumps at this period were primitive in the extreme: they were just barrels mounted on wheels or sleds, with a single-cylinder pump immersed in the water in the barrel and operated by long handles. A swivelling nozzle called the 'monitor' or 'gooseneck' above the cylinder permitted the water to be squirted at the flames, but the range was so limited that the engine had to be dragged as close as possible to the blaze, with the result that the engine itself often caught fire.

It was a Dutchman, Jan van der Heiden, born in 1637, who solved this problem. He invented a leather hosepipe, which enabled the firemen to take the hose right to the source of the fire. Van der Heiden, who was Superintendent of the Amsterdam Fire Brigade, was an accomplished engraver as well as an engineer, and in 1690 wrote and illustrated the first book entirely devoted to fire engines, entitled *Slang-Brand-Spuiten.* In it he described old and new methods of firefighting, showing how the introduction of hosepipe had revolutionized the science overnight. Before Van der Heiden's invention,

FIREMAN 1832
ROYAL EXCHANGE ASSURANCE

The elegant livery of some insurance brigades was surely designed for parades rather than practical firefighting

the typical Dutch fire engine was a metal cistern mounted on wooden skids, equipped with a twin cylinder pump, and fed with water from buckets brought by hand from the nearest canal. The fireman had to perch on top to control the direction of the monitor nozzle: altogether, the old pattern of engine was clumsy and inefficient.

In 1673, the Amsterdam Ropewalk – the 'Lijnbaan' – caught fire. Many of the buildings were a long way from the nearest water supply, and beyond the power of the old type of engine to control. But Van der Heiden's new pattern engine was brought into action for the first time: supplied with water through a flexible pipe fed by a portable canvas trough standing on a canal bank, and replenished by buckets dipped in the water, the engine was fitted with hose made in 50ft lengths, connected by brass screw joints. Because it did not need to hold so much water as the old bucket-fed pumps, Van der Heiden's engine was light enough to be carried by four men.

Now fires could be attacked at close quarters: the hose enabled firefighters to attack conflagrations that had been beyond the range of the old pumps – indoors, upstairs, or even on the rooftops. The success of the new engines at the Lijnbaan fire rendered the old pumps immediately obsolete, and they were replaced at once. Van der Heiden and his son were granted an exclusive manufacturing licence for the next quarter of a century: they further improved their invention by devising a suction hose, reinforced internally with wire, which could be dipped into a canal or pond to supply the engine with water.

But though Van der Heiden's invention of leather hose-pipe was to remain the standard equipment of fire brigades until the late nineteenth century – in 1818 Pennock & Sellers of Philadelphia improved the concept by patenting leather hose reinforced with copper rivets, which rendered it stronger and less liable to leakage than the sewn hose used previously – the Dutch did not retain their

On October 16, 1834, both Houses of Parliament were burned down: 12 engines attended, though the buildings were not insured

technological lead for long. 'Dutch engines' were introduced into England when William of Orange became King in 1688, and their design features were soon improved upon by the native manufacturers. By the early 1700s, the English had become the leading manufacturers of fire engines, and had even established an export trade – as early as 1679 an English maker had shipped a 'tub' hand pump to the American Colonies, for use in the city of Boston. King William didn't seem to benefit from the Dutch engines, for his London palace of Whitehall was twice seriously damaged by fire, firstly in 1691 'through the negligence of a maid-servant, who, to save the labour of cutting a candle from a pound, burnt it off, and threw the rest carelessly by before the flame was out'.

The growth of the English fire engine industry had, perhaps, been forced by a combination of legislation and cupidity: in 1707, the British Government made it obligatory for the churchwarden of every parish to make sure that all waterworks pipes were fitted with 'stop-blocks and fire-cocks', and the position of these clearly indicated by a mark fixed to the wall of the nearest building. Every parish, ran the law, had to provide a 'large engine', a 'hand engine' and a leather pipe and socket which could be coupled to the fire-socket. Failure to comply with these regulations meant a fine of £10.

The result: total apathy. More successful was an amendment to the law the following year, which promoted efficiency through the medium of thinly veiled bribes:

> The first person who brings in a parish engine, or any large engine with a socket when any fire happens, shall be paid 30s; the person who brings in the second, 20s; and the third, 10s.

With such inducements to hand it was not long before a British inventor developed a new kind of manual pump, which was to set the pattern for this type of machine for almost two centuries.

THE FIRST FIRE BRIGADES

Insurance firemen's uniforms from 1690 to 1860 and a Newsham-type manual, photographed in Birmingham in 1892

Richard Newsham, a London pearl button manufacturer, patented in 1721 'a new water engine for the quenching and extinguishing of fiers', in which the pistons were operated by chains running over toothed quadrants rocked up and down by the pumping levers: four years later, he patented another design which proved extremely popular and which was both widely used and, after his death in 1743, widely copied. Newsham had discovered a simple method of boosting the power output of the pump. In addition, he wrote, to the 'set of men to work at the levers as usual', Newsham's engine required 'a second set who stood on the engine above the levers, holding on to a handrail and stepping off and on to the treadles or footpieces as each lever was raised or depressed; their weight, in addition to the strength of the pumpers at the levers, giving the engine great power'.

Newsham manufactured his design in six different sizes; the No 5 Newsham was capable of throwing 160 gallons of water per minute to a height of 165ft. Because of their high silhouette and the handrails for the treadle operators, the Newsham engines were nicknamed 'four-posters'. Some of these parish pumps proved surprisingly long-lived: a Newsham engine purchased for the Devon borough of Dartmouth was still in use in 1875, and others were only pensioned off in the early part of the twentieth century.

In 1734 an author named Abel Switzer described Newsham's engines.

Richard Newsham, of Cloth-fair, London, engineer, makes a most useful, substantial and convenient engine for quenching fires, which makes continual streams with great force. He hath applied several of them before His Majesty and the nobility at St James's with so general an approbation that the largest was at the same time ordered for the use of the royal palace. The largest engine will go through a passage about three feet wide in working order, without taking off or putting on anything, and may be worked with ten men in the same passage. One man can quickly and with ease move the largest size about in the compass it stands in, and is to be play'd without rocking upon any uneven ground with hands and feet, or hands only, which cannot be paralleled by any other sort whatever. There is conveniency for about 20 men to apply their full strength and yet reserve both ends of the cistern clear from encumbrance, that others at the same time may be pouring in water which drains through large copper strainers.

The basis of Newsham's engine was a wooden trough waterproofed with pitch, which acted as a frame of the machine. The pumps were immersed in this trough, which could be filled either by bucket or by suction hose – a two-way cock was fitted to the inlet pipe – and water could be emitted either through a hose or a gooseneck monitor: the box containing the pumps also housed an air vessel to ensure a continuous discharge.

There was, however, one serious design defect: the Newsham engines were not particularly mobile. Designed for manual haulage, they had fixed axles, which made negotiating corners a major operation: in a confined space, ropes could be fixed to hooks at the corners of the frame to drag the machine round bodily. If a fire was at any distance, it was probably easier to lift the engine into a horse waggon.

Nevertheless, Newsham's 'four-posters' were superior to anything else that was available – one even fought a 'duel' against an engine built by John Fowkes of Westminster – and their fame soon spread to the English Colonies in the New World.

In 1658, when New York was still New Amsterdam, principal city of the Dutch colony of the New Netherlands, a fire brigade was organized, whose members, known as the 'Prowlers', were expected to walk the streets looking for fire from 9pm to 'morning drumbeat'. There were originally eight men in the brigade, also nicknamed the 'Rattlewatch' from the means of giving the alarm, but their numbers were soon swelled to 50. They were equipped with 250 leather fire buckets – sent from Holland – grappling hooks and small ladders.

Every citizen of New Amsterdam had to fill three buckets with water and leave them on his doorstep after sunset for the use of the fire patrol: additionally, ten buckets were to be filled at the town pump 'wen ye sun do go down' and left in a rack for the rattlewatch 'if ye fier does go further yan ye efforts of ye men and call for water'.

Once a fire had been extinguished, all the empty buckets were piled in a great heap on the common, and the town crier climbed on to a barrel and called out: 'Hear ye! O! I pray ye Lord! Masters claim your buckets!' This precipitated a great scrambling of small boys fighting for possession of private householders' buckets, for a rich citizen would usually give a small reward – a coin, a glass of wine, or a cake – for the safe return of his buckets.

By 1731, however, New York had grown to a city of 1200 houses, with 8628 inhabitants, and clearly some more reliable means of combating fire than the bucket patrols was needed. Therefore, on 6 May that year, the city authorities resolved 'with all convenient speed to procure two complete fire engines, with suctions and materials thereto belonging, for the public service; that the sizes thereof be of the fourth and sixth sizes of Mr Newsham's fire engines, and that Mr Mayor, Alderman Cruger, Alderman Rutgers and Alderman Roosevelt, or any of them, be a committee to agree with some proper merchant to send to London for the same by the first conveniency, and to report upon what terms the said fire engines, etc, will be delivered to the corporation'.

One Hundred Pounds Reward.

London, *June* 24, 1802

AS there is Reaſon to ſuſpect that many Fires have been occaſioned by the wilful Attempts of evil-minded Perſons, the *Governor* and *Company* of the *Royal-Exchange Aſſurance*, the *Managers* of the *Sun Fire-Office*, and the *Directors* of the *Phœnix Fire-Office*, do hereby offer a REWARD of

One Hundred Pounds,

To be paid on the Conviction of any Perſon, who ſhall, within the Term of One Year from the Date hereof, have wilfully and maliciouſly been the Occaſion of any Fire, which ſhall have happened in any Part of *GREAT BRITAIN.*

This Reward will be paid by either of the ſaid Offices, over and above all Parliamentary, Parochial, or any other, Rewards whatever.

By the ACT *of the 9th of* GEORGE I. *Chap.* 22, *it is enacted,*

That, if any Perſon or Perſons ſhall wilfully and maliciouſly ſet Fire to any HOUSE, BARN, or OUT-HOUSE, or to any HOVEL, COCK, MOW, or STACK of CORN, STRAW, HAY or WOOD, they ſhall be adjudged guilty of FELONY, and ſhall ſuffer DEATH without Benefit of Clergy.

Burning down one's own property to claim the insurance was a hanging offence in former times, as this 1802 poster proves

The Aldermen obviously did act with all convenient speed, for the engines were delivered the following December, and housed in the City Hall. Peter Rutgers, a brewer, and assistant alderman of the North Ward, was 'the first man that ever had charge of a fire engine on Manhattan Island'; the second was John Roosevelt, a merchant.

Nor did Rutgers and Roosevelt have to wait long to exercise their new responsibilities. On 6 January, 1732, the *Boston Newsletter* published a story under the prosaic heading 'News from New York': 'Last night, about twelve o'clock, a fire broke out in a joiner's house in this city. It began in the garret, where the people were all asleep, and burnt violently; but by the aid of the two fire engines, which came from London in the ship *Beaver*, the fire was extinguished after having burnt down the house and damaged the next'.

Soon, a proper fire department was organized: all able-bodied men over 21 were compelled to undertake fire duty, and the department was under the command of an 'overseer' named Anthony Lamb, who received a salary of £12 a year. He held the post until 1736 when he was succeeded by Jacob Turk, a gunsmith.

Soon, too, America had its own fire engine industry: probably the first fire engine ever built in the New World was 'Old Brass Back', a copy of the Newsham engine, manufactured in 1743 in the workshop of Thomas Lote, a cooper and boatbuilder of New York. Old

CORN DEALER AND HAY SALESMAN

'London Fire Engines – the noble protectors of lives & property' – horse-drawn manuals of the early 19th century

Manual engines rush to a fire at Blackfriars in the early 19th century

Brass Back, which survived for many years, was housed in a shed on the shores of Kalch-Hook Pond, a fishing lake in what was then rural New York.

There may have been an earlier American manufacturer, for on 9 May, 1737, the *New York Gazette* carried the following advertisement: 'A fire-engine, that will deliver two hogsheads of water in a minute, in a continual stream, is to be sold by William Lindsay, the maker thereof. Enquire at Fighting Cocks, next door to the Exchange Coffee-house, New York'. But whether Lindsay was a genuine manufacturer, or merely a merchant venturer, remains a mystery.

One early American manufacturer, Bartholemew Welden, is recorded as having built two engines in New York around this time, but as neither of the machines would work, his claim to fame is somewhat tenuous.

In 1737 the New York Volunteer Fire Brigade was formed under Jacob Turk's command, with a strength of 35 firemen: the truly volunteer nature of the unit was proven by the fact that the members paid for their two engines and equipment out of their own pockets.

Turk was succeeded in 1761 by Jacob Stoutenburgh, another gunsmith and one of the original 35 firemen; he commanded the brigade until the Revolution, when there were seven engines in service.

The ardour of this brigade was obviously superior to their efficiency, for in 1773 the following dispatch was sent to London:

On the 29th of December the Government-house in New York accidentally took fire, and so rapid was its progress that, in a few moments after the alarm, a thick cloud of fire and smoak pervaded the whole building, and in less than two hours it was entirely consumed.

The Governor's family (an unhappy maid-servant only excepted) was by the Divine Providence preserved from the flames; his daughter being reduced to the extremity of leaping out of a window in the second story, and her life saved by falling on a deep snow.

The flames were so rapid, that nothing but a small part of furniture of one room was saved, not even the Governor's commissions and instructions; and had it not been for the snow lodged in the roof of the house, joined to the effect of the fire-engines, most of the city of New-York would have probably been destroyed.

During the War of Independence in 1776, when the British occupied New York, the Volunteer Fire Brigade became a home guard, with Stoutenburgh reporting direct to George Washington: and when a small waterfront house built of clapboard and inhabited by 'dissolute characters' caught fire, it seems that the brigade was powerless to act. Fanned by the wind, the flames spread rapidly, and the entire city west of Broadway – 493 houses! 8 was completely destroyed.

Fire was still a major peril in 19th century New York, as this contemporary lithograph shows

To augment the efforts of the Volunteer Fire Brigade, the citizens of New York formed their own brigades in 1781 under such names as the Friendly Union, the Hand-in-Hand and the Hand-in-Heart fire Companies: wearing distinctive round hats with black brims and white crowns, the members of these brigades assisted at fires by removing furniture and effects from the burning houses. They were exempted from the more menial tasks like handling water buckets or working the engines.

A decade or so later, definite rules of procedure were laid down for the firemen of New York: the first man to reach the engine house after the alarm was given was to carry the leather hoses to the fire, while the next four took the 'bucket poles' (long poles each carrying twelve fire-buckets); the remainder of the crew were to haul the engine, 'bawling out and demanding the aid of citizens as they proceeded on'.

By 1805, the design of engines, and the size of the connecting screws for the hoses, were standardized in the city: in 1819, when there were over 40 fire engines in service, suction feed from the river or the water mains had almost completely replaced the old bucket chain.

Firefighting had still not attained the status of a science – indeed, the nature of fire itself was still not understood – and a glance at some British firefighting reports of a typical year in the late eighteenth century shows that luck still played a large part in the saving of property (with their background in insurance, Brigades were little concerned with the preservation of human life: it was usually a case of *sauve qui peut . . .*)

February 9th, 1774: This morning about half an hour past six o'clock, a fire broke out at Mr Wagstaff's, green-grocer, in James-street, Bedford-row, which intirely consumed the same; but by the timely assistance of Mr Brooks's engine, and the well-conducting the Foundling-hospital engine, the flames were prevented communicating any further, though the houses adjoining catched fire several times. One woman, who was a lodger, was burnt, and another jumping out of a two pair of stairs window was greatly hurt.

November 13, 1774: A fire broke out at the timber yard of Mr Flight, in Tabernacle-walk, Moorfields, which consumed all the timber and the floor-cloth warehouse in the same walk. The flames spread so rapidly that the London insurance engine was near being burned, and several of the firemen were terribly scorched in bringing it away.

But in that same year, a Building Act was passed by Parliament which compelled each London parish to keep three or more proper ladders . . . for assisting persons in houses on fire to escape therefrom'. Any churchwarden who failed to comply with these regulations was liable to a summary fine of £10.

FIREMEN OF LONDON AND NEW YORK

When the London Royal Exchange burned down in 1838, the chimes in its tower were playing the tune 'There's nae luck about the house'...

Up to the turn of the eighteenth century, there was no such thing as a full-time fireman: all brigades were organized on a part-time basis, with the men paid 'according to their deserts' whenever they were called out. The usual rate of pay seems to have been one shilling for the first hour, sixpence for each succeeding hour – and unlimited beer. Bystanders were usually pressed into service to man the pumps, and would work the handles with a rhythmic chant of 'Beer-Oh! Beer-Oh!', pumping and drinking until they were too duddled to do either. A typical nineteenth-century news story reported: 'Thomas Frith, labourer, was charged with being drunk and disorderly after assisting at the fire which burned down the Bull Inn at Shepperton . . .'

It was Napoleon who organized the first regular fire brigade, the Sapeurs-Pompiers of Paris, as a division of the French Army: but Paris had always been a fire-conscious city. In 1684 it was recorded that the Royal Library of Louis XIV at Paris was protected by a single-cylinder fire engine which could throw out water in one continuous jet to a great height; and in 1699 the Sun King granted an exclusive right to one DuMourier DuPerrier to build and maintain *'pompes portatifs'* to safeguard Paris. In 1722 there were 30 'royal engines' and probably as many more owned by the Hotel de Ville. Building on such strong foundations, the Sapeurs-Pompiers soon achieved international renown.

Britain's first properly organized municipal fire brigade – though its 80 firemen were still part-timers – was that of Edinburgh, in 1824. The men were chosen from masons, slaters, plumbers and other trades familiar with methods of building construction. To head this brigade, the citizens chose a full-time Master of Fire Engines, a 23-year-old surveyor named James Braidwood. Braidwood trained and drilled his motley brigade – composed mainly of the former insurance fire companies of Edinburgh – until they became the most efficient firefighting force in Britain.

'Manhattan', the typically elaborate manual engine of a 19th century New York volunteer brigade

He organized fire drills at 4 am, because most serious fires occurred at night, and in any case there were no gaping bystanders to get in the way of his men. He devised a signalling system using 30 different calls on a bo'sun's pipe, which could easily be heard above the uproar at a fire. He encouraged his men to fight the fire at its source, taking the hose into a burning building, creeping low to gain the benefit of the layer of relatively fresh air drawn in from outside by convection from the blaze. And he ordered his men never to enter a building alone, but always with a colleague in case one of the firemen was injured or overcome by fumes.

Braidwood crystallized his ideas and theories in a book on firefighting published in 1830, which received great currency at home and abroad; and when the leading London insurance brigades decided to amalgamate into a combined establishment in 1832, it was unanimously agreed that 'Mr James Braidwood, who has been Superintendent of the Fire Engines at Edinburgh for 7 years is to be appointed as the Superintendent of the General Fire Engine Establishment of London, at a salary of £250 per annum'.

The new London Fire Engine Establishment was composed of the brigades of the Alliance, Atlas, Globe, Imperial, London, Protector, Royal Exchange, Sun, Union and Westminster insurance companies; there were 19 fire-stations and 80 full-time firemen, nearly all ex-seamen. Conditions were almost monastically severe: two men were always on 24-hour watch, while the other members of the brigade had always to be within the station buildings in readiness for call-out at any time. Only minimal amounts of leave were granted – each man was officially on duty 164 hours a week, leaving just four hours spare time! Only sailors, it was felt, would put up with such long hours and the cramped living accommodation at the fire stations: and as the pay, over £1 a week, was well above that which could be earned on shipboard at that time, Braidwood

THE WHITE TURTLE.

RESPECTFULLY DEDICATED TO THE

A dramatic contemporary illustration – but probably not too far from the truth – of a race between 19th century New York hose cart companies to be first to reach a fire

'The Great Fire of February 8th, 1876, on Broadway', showing an early steam pump at work

soon had sufficient recruits for his force.

In 1834 the Houses of Parliament at Westminster caught fire, and were almost completely destroyed, despite the efforts of 12 fire engines and 64 men, aided at the pumps by detachments from the Brigade of Guards: only Westminster Hall was saved. As the buildings had not been insured, the proprietors of the London Fire Engine Establishment justifiably felt that they had been put upon, and petitioned Parliament to set up a properly organized firefighting service to replace the insurance companies and the old parish pumps. But the Government, scenting a considerable expenditure in the wind, took refuge in pomposity, replying airily that any interference on their behalf might 'relax those private and parochial exertions' which had hitherto afforded 'so much effect and so much satisfaction to the public'.

And so the insurance companies had to continue providing fire protection for central London; though the Government did issue firefighting apparatus to some of the larger public buildings. However, when the Tower of London Armoury was destroyed by fire in October 1841, it was found that the government-issued buckets, hand pumps and engines were in a woeful state of dilapidation.

At least the insurance companies kept their equipment in excellent order, though occasionally even they were powerless to act, as the day in January 1838 when the Royal Exchange caught fire. In those days there were no proper fire hydrants, just wooden plugs knocked into the watermains, which were made up from short lengths of hollowed-out tree trunk socketed together. Indeed, where no fire plug existed the firemen had to dig down to the main and bore a hole with a long spoon bit. Then either a standpipe was inserted or the water collected in a portable 'dam' into which the suction hose was dipped. At the best of times it was a messy business, and the firemen were usually soaked through: but January 1838 was exceptionally cold, and the fire plugs were frozen up. And when at last an alternative water supply was found, the engines froze solid, too. . .

Braidwood's brigade, though keen and well-trained, found themselves increasingly stretched to cope with the 600 or so major fires which occurred each year in London. The spread of trade and industry and the growing urban population multiplied the fire risk, and some of the more public-spirited docks and factories organized their own private brigades to help out the Fire Engine Establishment.

The most famous private brigade was undoubtedly that of Frederick Hodges, who owned a distillery at Lambeth and maintained two splendidly equipped manual pumps, which, though originally intended to protect the distillery against fire, also assisted Braidwood's men, accompanied always by the 'Fireman's Friend', a barrel painted bright red with gold-leaf hoops, full of Mr Hodges' famous gin, which even inspired poetry.

> If fire you want to put in,
> Try Hodges' Cordial & Gin;
> If fire you want to put out,
> Try Hodges' engine and spout.

ran a contemporary doggerel, which was duly inscribed on the side of an elaborately decorated manual fire engine presented to Hodges by public subscription, and which was displayed at the Great Exhibition of 1851 (where it was also part of the official firefighting arrangements).

Hodges claimed that his brigade was, in relation to its size, better equipped than the London Fire Engine Establishment: thanks to a specially built look-out mast 120ft high (and generous tips to cabmen who brought information of conflagrations), Hodges' men could often reach the scene of a fire before the Fire Engine Establishment, who thereby missed the reward paid to the first engine to the scene of a fire. 'Captain' Hodges was a trained engineer as well as a businessman, and realized the potential of steam fire pumps before they were adopted by the Fire Engine Establishment: indeed, long before the motor age, Hodges preferred to putter round London at the controls of his Merryweather steam carriage *Fly-by-Night* rather than ride in a horse-drawn brougham. Surprisingly, when he married at the age of 47 in 1877, Hodges retired to Munich, where he had studied hydraulics as a young man.

By now, the fire engine industry had become so specialized that it was dominated by a very few companies: and the undisputed leaders were Merryweather & Sons, of Greenwich, and Shand, Mason & Company, of Blackfriars, whose business and technical rivalry was to spur the development of the British fire engine up to the start of the twentieth century.

Shand, Mason, could trace their ancestry back to a mechanic named Philips who had started building Newsham-type engines in 1774, had been succeeded in 1798 by a Mr Hopwood. Twenty-two years later, the firm was taken over by W. J. Tilley, whose engines were both up-to-date in design and sound in construction – the little manual that he supplied to St Paul's Cathedral in 1839 was still in use 90 years later. In 1851 the company became Shand, Mason.

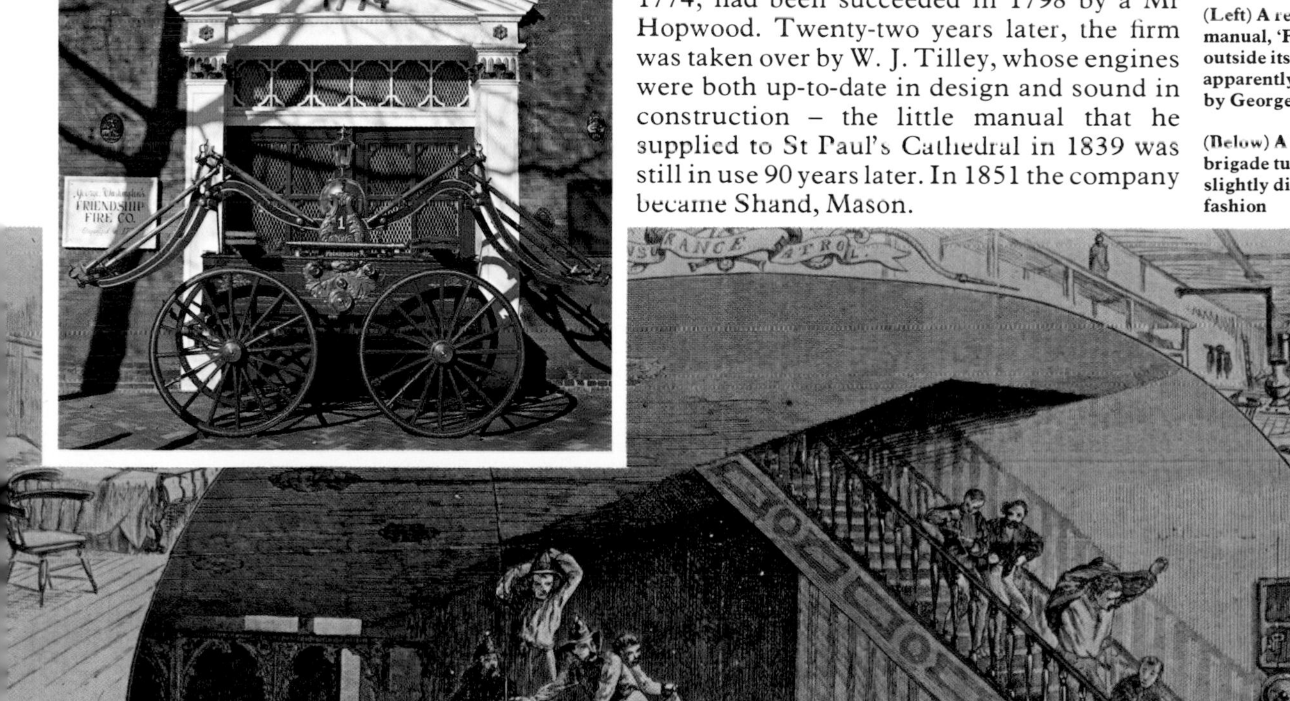

(Left) A restored American manual, 'Friendship' outside its firehouse, apparently founded in 1774 by George Washington

(Below) A New York brigade turning out in slightly disorganized fashion

Merryweather could trace their lineage back even further, to Samuel Hadley, who in 1769 had taken over a fire engine works, originally established by Adam Nuttall around 1750 in London's Long Acre. Samuel's son Nathaniel had joined the firm in 1790: two years later they took into partnership an engineer named Charles Simpkin, who had patented several improvements in engine design which became standard practice, such as valves of metal rather than leather, which were placed in separate valve chambers for greater accessibility. The short-lived leather valves of the earlier engines had been fitted inside the cylinders and airchamber, making maintenance difficult. Another innovation was the provision of a steerable forecarriage and road springs, so that the engine could now be towed to the fire by horses, rather than dragged by perspiring firemen.

Henry Lott, son of Squire Lott of Twyford Abbey, Berkshire, joined the company in 1791: married to Simpkin's widow, he later became a partner, and eventually took over the control of Hadley, Simpkin & Lott. In 1807 a 14-year-old apprentice, Moses Merryweather, a Yorkshireman related to Captain Cook, was taken on: in 1815 Henry Lott's niece was born. Young Moses was evidently an enterprising lad, for in 1836 he married the niece; Uncle Henry died soon after, and Moses took over the company.

Moses lived to the age of 79, though his son Richard headed Merryweather & Sons from 1859: on Richard's death in 1877, the company was taken over by his younger brother James Compton Merryweather, perhaps the most flamboyant fire engine enthusiast of the entire family.

Though its resources were stretched to the limit, Braidwood's Fire Engine Establishment was setting a pattern which would be followed by fire brigades all over the world. Its fame had even reached the ears of Queen Victoria, whose castle at Windsor was only protected by an obsolescent Simpson manual pump built on the Newsham pattern in 1820, and by the local parish pump. When, in 1851, a fire broke out at Windsor Castle while the Queen was in residence, she ordered a message to be transmitted *via* 'that interesting and most extraordinary apparatus, the electric telegraph' to Braidwood, requesting that his men should render assistance. Immediately a company of firemen was dispatched to Windsor aboard a special train hauled by one of Isambard Kingdom Brunel's broad gauge Great Western Railway engines.

Once horse traction became general, firemen were able to ride on the engine, and Braidwood designed a seating arrangement similar to that used on the contemporary knife-board horse-buses, with the firemen sitting back-to-back on a longitudinal bench on top of the machine. At horse-drawn speeds, it was a moderately safe seat, but as speeds increased when motor traction was introduced, the Braidwood body became definitely a 'siege perilous'.

Though he had been in so many ways an innovator, Braidwood was opposed to the introduction of steam-powered pumps. Steam was, of course, no new force – it had been applied to the propulsion of vehicles on road, rail and water for many years past, and steam pumps were a proven feature of the mining industry – but Braidwood thought that the more powerful steam pumps would encourage his men to revert to the old 'long-shot' method of fighting fires, and continued to put his faith in manual pumps. He did, however, introduce a floating steam fire engine to deal with waterfront fires, and a Shand horse-drawn steamer was being experimented with in 1861.

But that year a six-storey warehouse full of highly inflammable goods – hemp, cotton, saltpetre, sugar, tallow – in Tooley Street, by the Thames in Southwark, caught fire, apparently by spontaneous combustion, and was soon ablaze from floor to roof. Braidwood took personal command of the situation, and every available man and machine – even the Hodges brigade, aided by James Compton Merryweather – was pressed into service . . . but it proved powerless to control the blaze. Thousands of rats fled the buildings and jumped into the Thames; the flames spread into neighbouring warehouses, adding sulphur, tallow, oil and paint to the holocaust, and the streams of blazing liquid literally set the Thames on fire, igniting boats at their moorings and making the task of the fire-floats almost impossible.

Braidwood decided to investigate one of the burning buildings. Pausing to lend a friend whose eyes were affected by the smoke his red-spotted handkerchief, Braidwood had walked only a few paces when the red-hot brickwork of a warehouse wall suddenly bulged outwards and collapsed on to him: it was three days before his body could be recovered from the debris. London's first fire chief was given a fitting funeral: the cortege was a mile-and-a-half long, and all the church bells of the city tolled a farewell to the dead hero.

In New York, firefighting had developed along very different lines from London:

engines were maintained by volunt
brigades, whose activities were social a
political as well as directed towards the savi
of life and property. One good reason why th
was so was the legal requirement that all mal
citizens of a certain age should serve for
specified duration, either as a militiaman or as a volunteer. Since the militiaman's term of service was five years, and that of the fireman only three, there was obviously no shortage of volunteer firemen. Moreover, the life of a volunteer fireman was, though demanding and often dangerous, less closely organized than that of the militiaman, and more exciting. There were parades, rifle clubs and dining societies to enliven the off-duty hours: and, above all, there was a bitter rivalry between the various engine companies that engendered a fierce pride which sometimes spilled over into combat as each company sought to prove that it was the best.

Pride, too, was manifested in the elaborate paintwork and decoration applied to the engines. Typical was Engine No 13, *Eagle*, which was housed in a waterfront shed near New York's premier fish market. The Eagle Company had been founded in 1791, and in 1830 they took delivery of a new horse-drawn engine (which was usually pulled by four milk-white horses). Zophar Mills, President of the New York Fire Department, recorded the occasion:

> The company was mostly Quakers of the highest respectability. They were generally merchants and merchant's clerks. They had their new engine – the first one in this city that was silverplated, and probably the only one in this country whose brasswork was silverplated. The engine was painted black, gold-striped, highly polished, and the back had Jupiter hurling thunderbolts painted on it, in the best style of decoration. She was the most elegant engine ever seen in those days, and all this expense of decoration was paid for by the company and their friends.

Eagle's crew were fined heavily if they missed a fire, but in 1807, in the days of the old manually-drawn *Eagle*, at least one erring – and not so respectable – member of the brigade attempted to justify his absence from a fire and had his misdemeanour recorded in the company minutes: 'Harris Sage's excuse is received. He says at the time of the fire he was locked in someone's arms and could not hear the alarm'.

And, indeed, it was a matter of shame for a fireman to miss a fire call, for each brigade was convinced of the superiority of its own engine. Water hydrants appeared relatively

th
so
fro
eng
sepa
Th
pany
– to p
preven
reserv
instruc
by their
begin pu
'Play av
through hi
steady 60 u
though if it v
in danger of b
at twice the pac
being able to r
for anything up
ready to be repl
two. On one nev
it is recorded, a
pumping rate of
many were the c
crushed against t
when firemen jun
an engine was bei
As the water fr
splashed into the
keep up their he
one of you, now, w
foreman through
her lively, lads! Y
her sides in! Say y
But if the engin
ders or a more wil
slowly rise in th
full! She's up to th
And the engine
occasion the engi
engine hugged a

ne
of
ing

, on
of a
ping-
ambi-
ed that
ngineers
ay 12 and
ere notori-
oker's com-
phia-pattern
wo banks of
orking in two
nding on the
al because it was
with firearms for

lar fireman with
mayors of New
he volunteer fire
gious Tammany
M. Tweed, started
ft and corruption
of the Americus
in 1850. Tweed
, which organized
vote behind his
ne.
house on Henry
d to be the most

elaborately equipped in New York; and the tiger painted on the body of the company's machine came to symbolize the stranglehold that Tweed had on city politics.

Before 'the good old Tig--a-a-a-r' became New York's most famous fire engine, that distinction had belonged to the machine of No 33 Company – the *Black Joke.* Engine Company No 33 had started life in 1807 as the *Bombazula* Brigade, but when they moved into a new fire house on Henry and Gouverneur Streets in 1828, the local 'Lady Bountiful' offered them a golden fire-trumpet if they would rechristen their engine *Lady Gouverneur* in her honour.

But the men of the *Bombazula* Brigade were rough diamonds, mostly workers in the New York shipyards, and had little time for the Establishment, even when it held out golden trumpets. They took a vote on the proposed change of name, and decided that if they were going to take a new title, it might as well be one that had some more spirited associations. During the Anglo-American War of 1812, many of those who were now firemen had helped to build a privateer named *Black Joke,* whose piratical activities caused much discomfort to the English fleet, and whose commander lived near the engine-house. So *Black Joke* the company became, and the picture on their engine was of the little privateer.

The exploits of the *Black Joke* Brigade were legendary – but the seeds of their downfall were sown on Christmas Eve 1843, when a rival company, full of seasonal spirits, swore that they would take down the carved wooden eagle above the door of *Black Joke's* fire house.

The 'boys at the engine house', scenting trouble, laid an ambush, loading what was confidently referred to as a 'howitzer' with 'chains, bolts and slugs' to repel the invaders. In the event, however, this ultimate deterrent was not used, for one of the invaders had come equipped with a musket, and was about to discharge it at the *Black Joke* crew when one of No 33's men, Tom Primrose, knocked up the barrel of the gun. The flintlock fired, and the deflected musket-ball hurtled in through the open window of a doctor's house, almost carrying away the nose of the unfortunate owner, who was dozing peacefully in his bed at the time.

The sound of the shot aroused the local constabulary, who came running to disperse the brawl . . . but the enemies of the *Black Joke* would soon triumph. It was during the Presidential campaign of 1844 that a horseback parade of Whig supporters, headed by prizefighter Tom Hyer, was passing the *Black Joke* engine-house when someone gave a false fire alarm. No 33 Company, in an unthinking reflex action, rushed out – right into the thick of the procession, the *Black Joke* causing the Whig horses to rear and plunge, and reducing the parade to a state of hopeless chaos and confusion.

This coming on top of all the other misdemeanours of the engine company, was too much for the officials of New York City: and a few days later, *Black Joke* was ignominiously pushed into the Corporation Yard 'tongue first' as a sign that the No 33 Brigade had been disbanded. Not that they were the only ones to suffer this ignominy, for in 1847, it is reported, the Tomkins engine company was disbanded for mutiny: they had refused to pass water to another engine when ordered to do so by their foreman.

But the men of the old New York Volunteer Fire Department could be brave as well as quarrelsome – and perhaps their finest hours came one freezing cold night in December 1835, when New York was swept by a fire of unprecedented ferocity. It started in the store of Comstock & Adams in Market Street, a narrow thoroughfare lined with shops and warehouses, and within a short time had taken such a hold on the buildings and their contents as to be uncontrollable.

The task of the firemen was rendered virtually impossible, for the frost had penetrated the ground to such an extent that the fireplugs were frozen solid, and there was little water available for the engines. One of the first engine companies on the scene was No 13, the *Continental Eagle,* whose machine

Frederick Hodges's private fire brigade of the 1860s: 'Captain' Hodges stands proudly at the centre

(Right) The fire at Toppings Wharf, in London, on August 19, 1843, destroyed St. Olave's Church

(Opposite page) In 1862 this ornate Merryweather manual was presented to Frederick Hodges's private brigade by the parishioners of Southwark

had been the first in New York to be equipped with a suction hose.

Chief Engineer James Gulick and William H. Macy attempted to stop the fire from spreading – indeed, they did prevent the flames from crossing Wall and Water Streets – and were working so close to the blaze that Engine 13 was ordered to play its hose on Macy. Though the thermometer stood at 17 degrees below zero, Macy was in danger of being roasted alive. His coat was burned through and his leather hat 'roasted to a crisp': but he survived, and eventually became president of a savings bank for seamen.

Despite the water shortage, engine companies from New Jersey and Philadelphia raced to New York to assist the Volunteer Fire Department; one imperilled building was saved by the judicious use of vinegar when the water supply ran out.

Black Joke was dragged on board a ship at the foot of Wall Street, so that water could be taken from the unfrozen East River, and pumped into Engine 26, which transferred it to Engine 41, *Old Stag*, whose crew had taken novel precautions against the weather. 'Beside the engine stood a keg of brandy and a copper kettle', ran a contemporary report. 'Old firemen's eyes sparkle as they declare that the liquor was poured into the engine to keep her from freezing. It must have been poor brandy, for she froze in spite of it'.

Whether the brandy finished up inside the firemen or inside *Old Stag*, the engine was successfully thawed out and back in action the next morning.

At the height of the blaze a newspaper editor, Charles King, rowed across the East River to the naval yard to obtain gunpowder to create a fire-break around the blazing area, and sailed back in a brig manned by a volunteer crew of sailors and marines, with the kegs wrapped in blankets to prevent them being ignited by the flying sparks. Then, aided by his crew and the firemen, King blew up several large buildings to create a gap to the north-east of the flames across which the fire could not spread.

When the toll was taken, an area a quarter of a mile square was found to have been burned to the ground, consisting mainly of warehouses four and five stories high, and containing goods valued at between $20 million and $40 million.

Even though the elements were against them, the New York Fire Brigade had prevented the destruction from being even greater. In the plaudits that were handed out after the fire, the out-of-town brigades that had sped to the rescue were not forgotten: typically, the Franklin Engine Company of Philadelphia were presented with a specially commissioned back panel for their engine bearing a portrait of Benjamin Franklin at his printing press.

The New York Volunteer Brigade survived until 1865: in America at large, the volunteer brigade concept still survives.

1828: THE STEAM ENGINE ARRIVES

One of the earliest Merryweather steam pumpers in service in Holland in the 1860s

Though steam was the driving force behind the Industrial Revolution – and had indeed first been applied to pumping engines, albeit of monstrous size, in the Cornish tin mines – fire engine makers (and users) were, for an inordinately long time, chary of using steam to fight fire. As late as 1860, John Decker, chief of the New York Fire Brigade, could pontificate:

> At large fires steam engines are serviceable auxiliaries to the hand engines, but they can never take the place of steam apparatus, as eight fires out of every ten that occur are brought under subjection by the quickness of operation of the hand engines, so that there is no necessity for placing the steamers to work.

It was a point of view that had persisted, sometimes more vehemently expressed, for over 30 years, ever since that brilliant team of inventors George Braithwaite and John Ericsson, of New Road, London, had built the world's first steam fire engine in 1828–9. Of remarkably advanced concept, it had a 10 h.p. engine with two horizontal cylinders of 7in bore in tandem with pumps of 6½in bore, which developed sufficient pressure to throw 150 gallons of water power a minute to a height of some 90 feet. It could reach working pressure within twenty minutes of the boiler being fired, and the waste steam from the cylinders was led through the feed water tank in two coiled pipes, thus raising the water's temperature considerably before it was pumped into the boiler. The firebox was water-jacketed, and fanned by a forced draught from a mechanical bellows.

Braithwaite and Ericsson soon had a chance to show what their machine would do: the celebrated Argyll Rooms in London caught fire in the depths of winter, and the cold was so intense that all the manual engines which attended the blaze froze up. But the Braithwaite engine kept going for five hours, throwing a continuous stream of water right up to the dome of the building.

It was an undoubted success, but firefighting opinion was against it, claiming that the engine was too powerful for the street mains to supply, and encouraged the press to pillory 'Braithwaite's kitchen stove', condemning it as a mere 'steam squirt'.

Curiously, Braithwaite and Ericsson relied on horse traction to take their engine to the scene of the fire; curiously, for the same year that their fire engine appeared, they entered a railway engine of similar overall design for the famous Rainhill Trials organized by the Liverpool & Manchester Railway (which was won by Stephenson's *Rocket*). Their railway engine *Novelty* was the fastest of the four contestants, with a reported top speed of 28 m.p.h., but it was mechanically frail, and the failure of first the mechanical bellows and then the feed pipe from the boiler to the cylinders eliminated it from the contest.

Bad luck dogged their venture into fire engines, too, for there were few people willing to order such a controversial device. When they helped to put out a fire at Barclay's Brewery, the brewers asked to borrow the engine for a month: but to pump beer, not water, which it did night and day for the whole period.

Braithwaite and Ericsson built another fire engine two years later, a smaller machine with a single-cylinder engine, while a more powerful development of their original design, with a geared-up, three-barrel pump, was ordered for Liverpool. Royal patronage came in 1832, when the King of Prussia commissioned an engine from the partners: known as *Comet*, it had the two-cylinder, two-barrel pump layout, was rated at 15 h.p., and could discharge 300 gallons of water a minute.

A fifth engine was built in 1832, but it was their last: Ericsson, born in Sweden in 1803, subsequently went to America, where he achieved fame by designing the iron-clad battleship *Monitor* for the Northern States in the Civil War. Armed with 11-inch guns, *Monitor* was the first-ever ship to have a revolving gun turret. Ericsson's entire career was a succession of triumphs and failures – at one stage he was imprisoned in the Marshalsea debtors' prison in London – but before his death in 1889, aged 86, he was to see his original concept of the steam fire engine become generally accepted. Another of his inventions was a hot-air – 'caloric' – engine, a type of power unit which is currently being investigated as an alternative to the petrol engine.

Next to attempt to build a steam fire engine was the firm of G. Rennie & Sons, who supplied a steamer to the Woolwich Naval Dockyard in 1836, though they hedged their bets by constructing a dual-purpose machine, which could also be used as a pump for emptying the dockyard caissons.

America's first steam fire engine was built in New York in 1840 by Paul Rapsey Hodge, an Englishman born in St Austell, Cornwall, in 1808.

The engine had been commissioned for trials by insurance companies, and proved to be a remarkable machine in many ways. Work started on 12 December, 1840, and the machine was ready for trial on 25 April the following year. The framing and wheels were built by the Matteawan Company to Hodge's design, and the Bury-type boiler followed contemporary railway practice. The two cylinders, each with a pump cylinder in tandem, were mounted outside the framing, and there was a tall air vessel ahead of the boiler, surmounted by a brass bell and an eagle. Long coupling rods linked the cylinders to the rear wheels, making this the first-ever self-propelling fire engine – and in its general lines it presaged the design of the traction engines which would appear for the first time towards the end of the decade.

There was no way of throwing the wheels

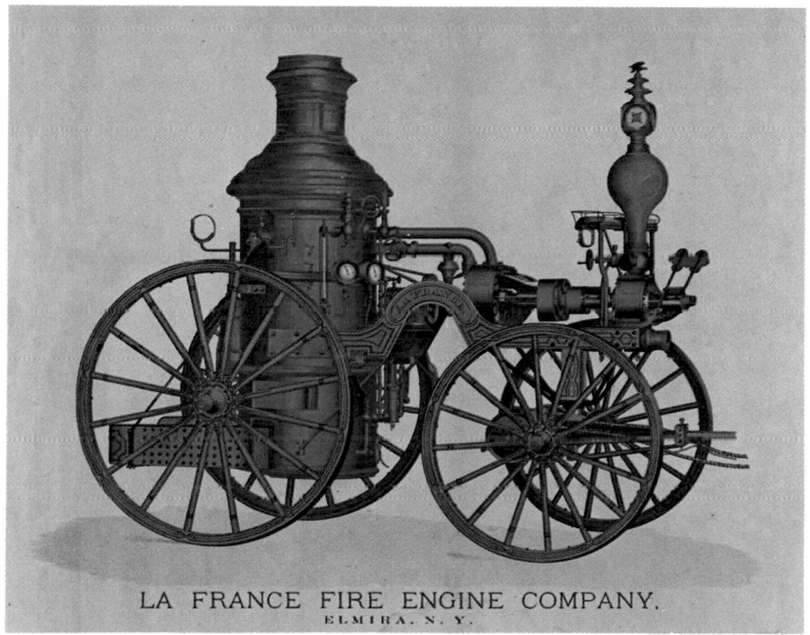

(Top) 'The Great Engine Contest of July 1st, 1850' between the engines of the Philadelphia Fire Department

(Above) Truckson La France began building engines of this type in the 1870s in Elmira, New York State

out of gear, so that when the engine reached the fire, it had to be jacked up until the rear wheels were clear of the ground, when they spun round and acted as flywheels.

Though Hodge's engine was self-propelling, it could move no faster than the horses between its shafts – and if that seems like a contradiction in terms, the simple answer is that the horses weren't there to pull (though they might have acted as an emergency 'low gear' on hills) but to steer the engine! Many of the very early traction engines had horse steering (though why it was beyond the wit of these engineers to devise a steering mechanism when the steam carriages of the 1820–40 era had had them all along is an unanswerable question), and it was said that a horse which had been employed for steering was henceforth spoiled for normal draught work when it actually had to pull its load.

Like the Braithwaite and Ericsson engine, Hodge's machine was greeted with scorn and derision by the firefighting authorities, and its development was not proceeded with: in 1850 Hodge returned to England.

It is interesting to note that this prejudice against steam fire engines was so pronounced that when in 1842 Moses Merryweather was asked to produce a fire engine to run on the London & Birmingham Railway, a manual was specified. But what a manual! Requiring a pumping crew of 42 men, it was the largest hand-pump engine that had ever been produced. Fitted with flanged wheels so that it could be drawn to any point on the company's network, it had spring buffers that could be dropped down out of the way of the pumping crew. It carried 450 gallons of water in a cistern 13 feet long, and its gun-metal cylinders were 9 inches in diameter and 10 inches long. A hose 400 feet long was carried on a reel at the front of the engine, and with a 1¼-inch diameter nozzle, a jet of water could be thrown to a height of over 100 feet. Increasing the pressure by restricting the nozzle aperture gave an even greater range.

Another ingenious self-propelled fire engine appeared in 1851 in Philadelphia. Designed by one W. L. Lay, it could be driven to the fire by 'carbonic acid gas' while steam was being raised.

In 1852 the Cincinnati, Ohio, council having offered a prize of $5000 for a successful steam fire engine, a sharp character named Alexander B. Latta claimed to have perfected such a unit (though the machine afterwards proved to have been designed by Latta's partner, Abel Shawk). Called *Uncle Joe Ross* in honour of a member of the council, the Shawk-Latta engine was a massive three-wheeler scaling 22,000 lb, and ponderously self-propelling (though if contemporary engravings are anything to go by, its driving mechanism was woefully inadequate for the machine's bulk). Four horses were needed to steer the single front wheel. This engine could, using a 3-inch hose and a 1½-inch nozzle, project water for 225 feet, 'to the astonishment of thousands of people present'.

In 1855 Shawk built a second machine under his own name. The A. B. Latta Company survived for another decade, then in 1862 was sold to Lane and Bodley; in 1868 it was purchased by an employee named Chris Ahrens, who improved the design – and changed the corporate name to Ahrens Manufacturing Company.

By the middle of the nineteenth century, America could boast a number of well-established indigenous fire engine makers. One of the best-known dated back to 1832, when one John Rogers had begun manufacture of hand-pumped engines beside King's Canal, in Waterford, New York. Before long, the company had been taken over by a 23-year-old employee, Lysander Button, a Connecticut Yankee from New Haven. In the fullness of time, Lysander was joined by his son Theodore, and the expanding company moved to a steam-operated plant on the Erie Barge Canal, where they built hand engines in sizes ranging from four-man-operated to 60-man-operated. Boasted the Button catalogue:

> As the years go by, the reputation of our engines has increased so that they are admitted to be the standard Hand Fire Engine of the age. It is not egotistical to say that no substantial improvements have been made in the construction of fire engines during the last fifty years that have not been copied from or made in imitation of the Button engine.

In 1862 the Button Fire Engine Company began building steam fire engines: the first one was sold to Battle Creek, Michigan, and by 1896 Button engines had earned a considerable reputation:

> They were received with great enthusiasm by firemen everywhere . . . their position as producers is established over the entire globe. It is, indeed, a great honour, but laurels fall gracefully, and there are none who should be envious enough to say they have been misplaced. These engines have taken more than two-thirds of all prizes at musters and challenge tests in the last twenty-five years.

But there were other American makers, too: and an increasing number of them seemed to be settling in the area of one little village, Seneca Falls, 60 miles north of Elmira, in New York State, which had the geographical advantage of being located on the New York State Barge Canal. The first manufacturer to set up business in Seneca Falls was Paine & Caldwell in 1839. In 1845 came the Silsby Manufacturing Company, who became famous for the manufacture of steam fire engines – in 1861 they supplied a hand-drawn steamer with a 'rotary engine' to the New York Fire

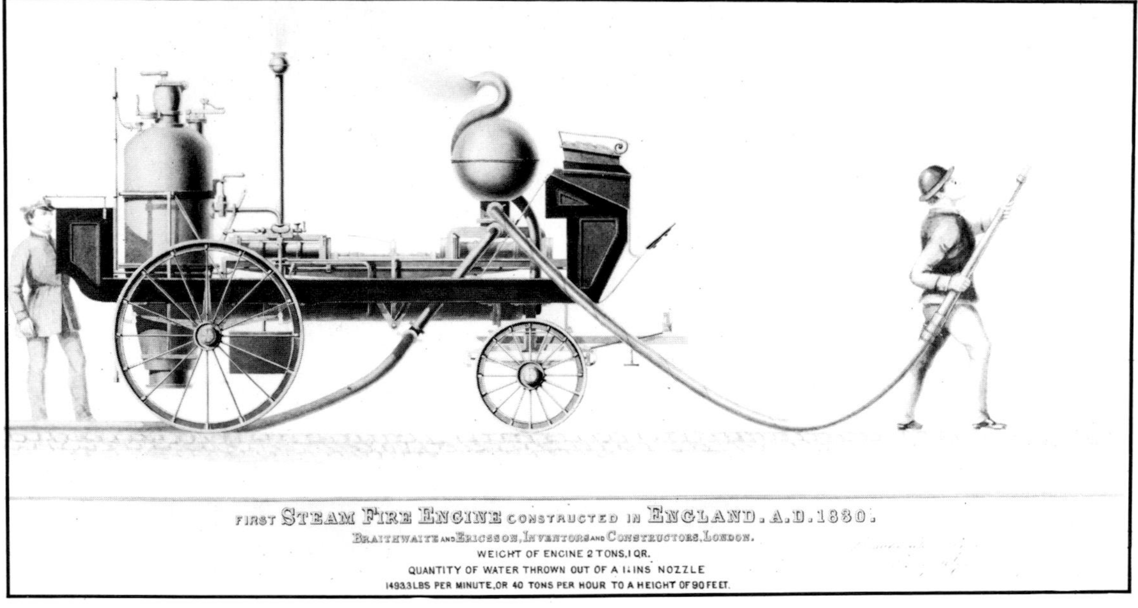

FIRST STEAM FIRE ENGINE CONSTRUCTED IN ENGLAND. A.D. 1830.
BRAITHWAITE AND ERICSSON, INVENTORS AND CONSTRUCTORS, LONDON.
WEIGHT OF ENGINE 2 TONS, 1 QR.
QUANTITY OF WATER THROWN OUT OF A 1¼ INS NOZZLE
1493.3 LBS PER MINUTE, OR 40 TONS PER HOUR TO A HEIGHT OF 90 FEET.

(Above left) The world's first fire engine, built by Braithwaite & Ericsson in 1829-30: though entirely practicable, it was condemned by popular opinion

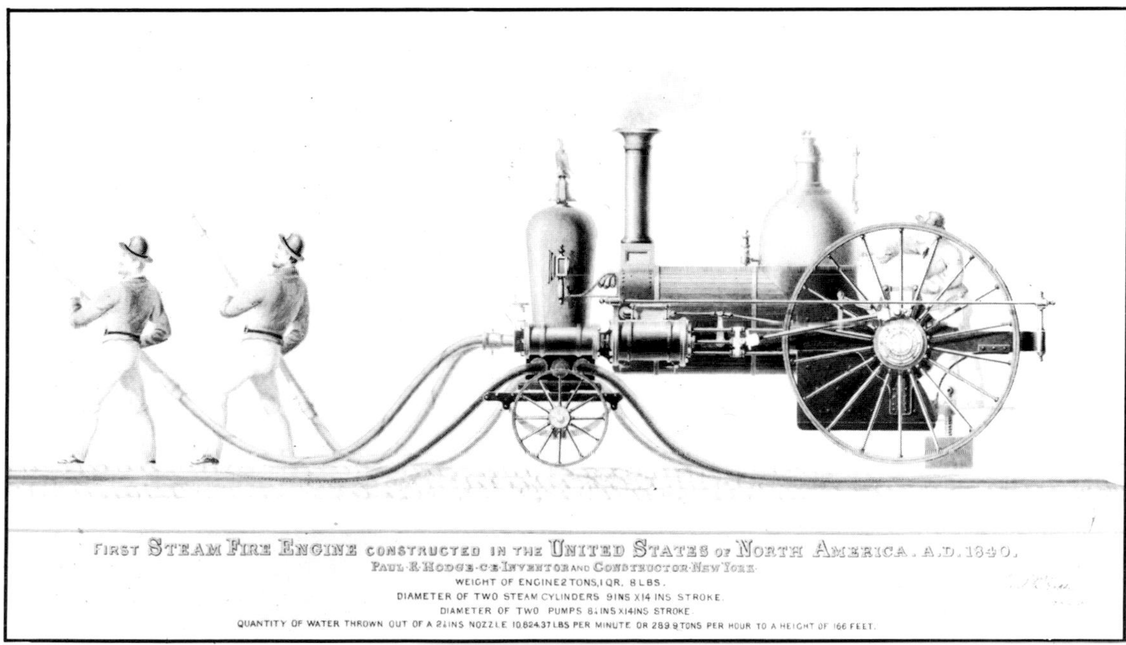

FIRST STEAM FIRE ENGINE CONSTRUCTED IN THE UNITED STATES OF NORTH AMERICA. A.D. 1840.
PAUL R. HODGE C.E. INVENTOR AND CONSTRUCTOR NEW YORK
WEIGHT OF ENGINE 2 TONS, 1 QR. 8 LBS.
DIAMETER OF TWO STEAM CYLINDERS 9 INS X 14 INS STROKE.
DIAMETER OF TWO PUMPS 8¼ INS X 14 INS STROKE.
QUANTITY OF WATER THROWN OUT OF A 2¼ INS NOZZLE 10,824.37 LBS PER MINUTE OR 289.9 TONS PER HOUR TO A HEIGHT OF 166 FEET.

An Englishman, Paul Rapsey Hodge, built America's first steam fire engine (below left) in 1840 in New York

Brigade. Other companies in the firefighting business who settled in Seneca Falls were Gleason & Bailey (1884), who made hose wagons and hand pumps, Rumsey & Company (1864), builders of hose carts and ladder trucks, and the American Fire Engine Company (1891).

There were so many firefighting factories in Seneca Falls that it eventually became known as 'The Fire Engine Capital of the World'.

Philadelphia, too, had a number of fire engine makers, most of whom were represented at the great trial of steam fire engines organized in that city in 1859, which was probably the first real public demonstration of the state of the art of building and operating steam fire appliances. The engines shown were *Good Intent, Hibernia* and *Mechanic,* built by Reany & Neapy of Philadelphia, *Independence,* constructed by Hunsworth, Eakins & Company, of the People's Works, Philadelphia, Merrick & Sons' *Weccacoe,* also from the City of Brotherly Love, as were the *Assistance* and the *Philadelphia,* whose makers were not recorded. Poole & Hunt of Baltimore, founded in 1858, were to build only seven engines in eight years, and of these two were at the Philadelphia Trials, the *Baltimore* and the *Washington,* of which it was said that there was 'no screw, bolt, or handle or any other appurtenance more than necessary'. Other engines attending were the *Citizen* of Harrisburg, the *Globe* and the *Franklin* (of Frankford), and the *Southwark,* built by Lee & Larned, of New York.

That same year Lee & Larned supplied two of their engines to the New York brigade, which used them at the Duane Street fire on 17 January and the South Street Fire on 24 January. Fire Chief Harry Howard was dubious of their worth:

> The steam engines were both large in size and powerful in action, and if permitted to discharge water at every fire would entail more damage by that element than the one

(Right) Two manual engines of the London Salvage Corps, still in service in the early 1900s

it was sought to subdue. The propriety of their introduction was questionable in my judgment, though their services might be rendered effective on extraordinary occasions when the department might be called on to do extra or laborious duty.

But when Lee & Larned sent one of their engines to the International Exhibition in London in 1862, it was reported that it 'elicited the greatest approbations from all the engineers who inspected it, both for its beautiful workmanship and satisfactory working'; Lee & Larned actually built a small engine in England that year.

There were other steamers in service in New York at the end of the 1850s: in 1859, James Smith, who built fire apparatus at his works in West Broadway, built a 'piston engine of small size', intended to be drawn by hand, which was used with great success by the Hudson Hose Company No 21, while in 1860 the Amoskeag Manufacturing Company of Manchester, New Hampshire, brought a steam fire engine to New York to prove its paces. 'It was afterwards bought, and proved to be a valuable auxiliary to the fire service'.

More steamers were bought in 1861, from the Portland Manufacturing Company, from A. B. Taylor & Son, of New York ('James Smith's Patent'), one of which was sold to the revived Black Joke Company, from Jos. Banks, Mr Jeffers and A. van Ness, all of New York City. Up to 1865, when the Volunteer Fire Department was succeeded by the New York Metropolitan Fire Department, these engines were all hand-hauled, but thereafter many were converted for horse traction.

Belief in the superiority of man power over horse power was an unconscionable time a-dying: mistrust of horse traction dated back to the 1820s, when Hook & Ladder Company No 1 had been the first New York brigade to get a horse. Not only did the manually hauled hook and ladder truck of No 11 unit beat them to the fire, but the horse was so badly winded that it was never any use afterwards!

In Europe, horse traction was used from the start: the first successful manufacturer of steam fire engines was Shand Mason, whose first land steam fire engine was built in 1856, and exported to Russia: but the London Fire Brigade did not order a steamer until 1860. In July that year, a massive Shand Mason machine was used in one of the back streets of Doctor's Commons: it weighed 84 cwt and needed three horses to draw it, despite which Braidwood remarked acidly that the 'land steamer . . . required delicate handling'.

Ireland's first steam fire engine appeared the same year; it was built by James Shekleton, an engineer from Dundalk.

By that time, the first European self-propellor had made its debut: it was the *Steam Elephant,* a maid-of-all-work machine built in 1859 by James Taylor, a traction engine manufacturer of the Britannia Works, Birkenhead, Cheshire, and seems to have been the first self-propelled fire engine capable of steering itself without the aid of horses. It was also designed so that one man could attend to the tasks of driving and stoking the engine single-handed, unlike the majority of contemporary designs which needed both steersman and stoker.

The magazine *The Artizan* waxed eloquent over the manifold virtues of the *Steam Elephant:*

Now Mr Taylor, who is well-known for the admirable steam winches, cranes, hoists, and such-like labour-saving machines invented and constructed by him, has for some time past devoted himself to design-

ing a portable steam engine, which shall be capable of running over ground of variable degrees of hardness, drawing loaded trucks or agricultural implements, and also for performing the duties of an ordinary portable steam engine for driving or rotating machinery, for raising and lowering heavy weights and, by the application of a derrick or sheer leg, to perform the duties of a crane, besides containing within itself the means of performing various other descriptions of work, as that of a crab, winch, windlass, etc.; it may be employed for pumping water or as a fire engine.

It seems, though, that the customers were mistrustful of this Jack-of-all-trades versatility, for though a revised *Steam Elephant* appeared at the 1862 Exhibition, the type then faded into obscurity.

Another self-propellor, this time designed specifically as a fire engine, appeared in 1861, the design of William Roberts, of Millwall, London. A three-wheeler of somewhat uncouth appearance, Roberts's engine both steered and drove with its single front wheel. Just 12 ft 6 ins long and 6 ft 4 ins wide, the unwieldy brute scaled an earthquaking 7½ tons, yet was capable at running at 18 m.p.h. (though at what cost to the road surface contemporary reports were tactful enough not to mention!).

Under test, Roberts's engine showed itself capable of throwing a jet 140 feet high over a horizontal distance of 182 feet. In 1862, Roberts built a second engine, named *Princess of Wales*, in honour of the 'fairy on England's Christmas tree', the immensely popular Princess Alexandra, who had just married Prince Edward, the heir to the throne. This seems to have been a machine of greater dimensions, for it is reported that it could accommodate eighteen men, 'with a great

Frederick Hodges bought Merryweather's first two production steam fire engines for his private brigade: Torrent **was built in 1862**

An elaborate fire insurance certificate of 1860: among the trustees of this New York company was John Jacob Astor

quantity of appliances, ladders, etc.'.

In a trial of steam fire engines held at the Crystal Palace in July 1863, *Princess of Wales* showed itself capable of getting up working steam pressure within twelve minutes of the boiler being fired.

The third Roberts steam fire engine, *Excelsior*, was built in 1865, and this strange device was shipped to Rio de Janiero, where it was used in that city's arsenal, surely the last place that the new concept of fighting fire with fire should have been applied . . .

But by that time there was small point in building any form of self-propelled vehicle for use on British roads, for in 1865 the ridiculous Locomotive Act (which was in part inspired by the road-destroying activities of behemoths like the *Steam Elephant*), had restricted the velocity of road locomotives to four miles an hour on the open road and two miles an hour in populated areas, and compelled them to be in the charge of 'at least three persons . . . one of such persons . . . shall precede such Locomotive on foot by not less than Sixty Yards, and shall carry a Red Flag constantly displayed.'

Horse-drawn vehicles, apparently, were exempt from these lethargic speed limits (as were electric tramcars): so the British self-propelled steam fire engine had to wait until the law became more amenable.

But the horse-drawn steam engine now entered on an era of rapid development – apparently because the volunteer pumpers kept striking for more beer! – with manufacturers vying with each other to produce new designs. Shand Mason had taken the initial advantage, with engines which could pump at 200 strokes per minute, faster than the fastest manual: one of these, on trial at Waterloo Bridge, succeeded in throwing a jet to a height of 140 feet.

Then, in 1861, Merryweather joined the fray, with the *Deluge*, whose massive 30 h.p. single cylinder engine had a bore of 9 inches and a stroke of 15 inches, and drove an outsize twin-cylinder pump, which had bores of 16½ inches and strokes of 15 inches. It could develop more pressure than the Shand Mason engine, and was capable of hurling a far more powerful jet clear over the top of a 140 ft-high chimney.

Costing £700, *Deluge* won the large-engine class in Britain's first-ever trial of steam fire engines, held in London's Hyde Park in conjunction with the Great Exhibition of 1862. A Shand Mason won the small-engine class.

Deluge was acquired by that devoted amateur Frederick Hodges for his private fire brigade; later, at the time of the Franco-Prussian War in 1870, it was sold to the City of Lyons.

A second Merryweather steamer emerged from a newly-acquired factory in Lambeth in 1862, and was christened *Torrent;* a more powerful engine, the *Sutherland,* was introduced in 1863, with twin steam cylinders, and capable of projecting a steady jet 160–170-ft high through a 1½-inch nozzle. Its performance was so impressive that it won first prize for large steam engines at the 1863 Crystal Palace competition, and was subsequently purchased by the Admiralty, who placed it in service in Devonport Naval Dockyard, where it remained in commission for 27 years, and was finally taken off the active list in 1905, when it was replaced by a Merryweather *Greenwich Gem* steamer. It was then removed to Merryweather's museum, but 1918 saw it back in service, assisting a works boiler to meet an exceptional demand for steam power. Finally, in 1924, it was presented to the London Science Museum, where it remains to this day, almost certainly the oldest surviving steam fire engine in the world.

Another firm of engineers in Greenwich, T. W. Cowan (who also built a self-propelled private steam carriage about this time), constructed a special steam fire engine for the Tsar of Russia in 1863. Drawn by three horses and weighing 4 tons, this engine was intended to protect the Royal Library at St Petersburg against fire.

Also in 1863, a Herr Egestorff built a steam fire engine for the German city of Hanover.

The world leaders, it seemed, were still Shand Mason, for while all these other engineers were building in penny numbers, Shand Mason had already built up a thriving and prosperous export trade. Their output for 1863 totalled 17 steam fire engines, of which two – the new *London Fire Brigade Vertical* model – were for the London Fire Brigade, two were for Lisbon, three were for the Bombay and Baroda Railway Company (perhaps these were similar to those the company had built for Britain's London & North Western Railway two years previously, which were equipped with flanged railway wheels), four for Russia, two for New Zealand, one for Austria, one for Poland, one for Denmark and finally one for Dublin.

But by 1865, Merryweather had managed to increase their output to eleven steamers: and between them, Merryweather and Shand Mason were to determine the overall design for all subsequent British steam fire engines. Their rivalry was intense – at one trial, the Merryweather was unable to raise steam, then the engineer found that the Shand Mason's crew had stuffed oily rags down his chimney overnight!

One reason for the increase in demand was that the man who had taken command of the London Fire Engine Establishment after Braidwood's death was an enthusiast for steam fire engines – he was Captain Eyre Massey Shaw, an ebullient giant of an Irishman who had previously been Chief Constable and Chief Fire Officer of Belfast. He came to power at a particularly difficult time for the Fire Engine Establishment: faced with a £2 million bill for the Tooley Street conflagration, the insurance companies had been forced to raise their premiums 300 per cent, which occasioned a mass protest by the City of London merchants to the Lord Mayor. Conceding a small reduction in premiums, the insurance companies informed the Home Secretary that the cost of maintaining their firefighting establishment was too great after all, they received no financial assistance from any public body – and that they proposed to disband the force.

Faced with the prospect of having no fire service at all in London, the Government reacted quickly, and a Royal Commission was set up to investigate the 'Existing State of Legislation and any existing Arrangements for the Protection of Life and Property in the Metropolis'.

After deliberating for six weeks, they suggested that a fire brigade should be formed as a division of the Metropolitan Police: Shaw estimated the cost of such a scheme at £70,000 a year, but the Government said that this was too much, and lopped £20,000 from the total.

Though police brigades were operated by some smaller authorities, it was found to be impracticable to organize a police brigade for so huge and diverse a city as London, and in 1865 the Metropolitan Fire Brigade Act was passed, making the Board of Works responsible for the organization of London's fire protection.

On 1 January, 1866, the Metropolitan Fire Brigade officially came into being: it had a genuine baptism of fire, for on the first day of its existence, the Brigade had to deal with a major fire at St Katherine's Dock, in which property worth £200,000 was destroyed. Malicious critics claimed that the Brigade had only half-heartedly dealt with the blaze: though this was demonstrably untrue, it was not the last time that Massey Shaw's brigade would be the centre of controversy.

AMERICAN INNOVATIONS

MORTON
HOUSE
THE
BROADWAY
CAFÉ.
SURANCE.
850
DERMASALVE.
DR. SCOTT'S
CURES ALL DISEASES OF THE SKIN
STAR TH
DR. SCOTT'S ELEC
846 HERMON ATWOOD. 846
THEATRE
PARLOR.

While little progress was made in the development of self-propelled fire engines in Britain during the latter half of the nineteenth century, American inventors had their hands forced by an epidemic – the 'epizoötic' disease – which raged across America, wiping out horses in droves.

Curiously enough, one of the first self-propelled fire engines to run in the streets of New York since Hodge's brave venture in the 1840s was not propelled by steam, but by the novel internal combustion engine. It was devised by a young engineer called Reuben Plass, whose father ran a gas engineering and general machine shop on East 29th Street. With four massive cylinders cast on similar lines to the cannon the company had made during the Civil War, driving a fifth wheel through a sliding gear change, Plass's engine was ingenious rather than elegant. It seemed to answer its handlebar steering well enough, but young Reuben had omitted to provide for any sort of braking system in the design. Not unnaturally, the engine crashed spectacularly *en route* to a fire, and was promptly forbidden the streets of New York by the irate Mayor. Plass sold the remains to an engineering company in Cincinnati.

The early self-propelled steamers were, it seems, no more trustworthy: in 1878, the New York Engine Company No 5, based in Brooklyn, were given a new self-propelled engine. The local roads were badly surfaced and jolted the machine about, so the fireman in charge decided to take his unwieldy charge onto the tramlines, where the going was smoother. But the wheels engaged themselves inextricably in the tracks, and the engine ran away downhill and rammed a horse-tram.

After clearing up the damage, the district engineer took over the command of the fire engine. 'I got along pretty well until I was going up Atlantic Street', he commented laconically. 'Then I came behind a junk-wagon, which was filled with bottles, standing in front of a saloon. The propellor was making a great deal of noise; the old horse looked around, spied us coming up, and, without stopping to see if we had any licence, made a beeline up the street, with bottles flying like a volcano'.

Looking for somewhere quieter to test the machine, the engineer decided to try its hill-climbing powers, and his steersman headed for Fulton Street, where there was a suitable gradient. At this point, the man at the helm seems to have become hopelessly flustered – maybe it was the strain of handling such a large machine – for the unfortunate engineer suddenly lost all realization of what was happening until he found himself sprawled against the boiler, which he was hugging like a long-lost brother, with the hot metal burning the tip of his nose.

Apparently, the driver had allowed the fire engine to swerve up on to the pavement, where it had uprooted two iron posts, and only been saved from demolishing a house when the front wheels had caught in the gutter.

Covering up the evidence of the fire engine's progress proved to be a simple matter of finance, and the unfortunate engineer had to hand over $20 to repair the damage, which, under the circumstances, must have seemed a mercifully cheap deliverance! The machine was handed over to another engine company to see if they could fare any better.

When they read of such alarming goings-on, those who could still get horses to pull their steam engines must have counted themselves fortunate. It was the proud boast of the leading fire brigades that they could have their horses harnessed into the shafts – 'quick hitch' collars were suspended from the ceiling – and ready for action within ten seconds of receiving the call to duty: one company even claimed that it could be ready for action in just six seconds . . .

The manufacture of steam fire engines was now becoming an attractively profitable business, (though both in Britain and America the decline of the manual pump was long and drawn out, for many small brigades, especially those attached to large houses or private companies, found that the manual was more suitable to their needs and pocket), and some of the great names of the industry were being established.

One of the more remarkable success stories was that of a young iron-founder named Truckson Slocum LaFrance, who one day in the 1860s arrived in the little township of Elmira, New York State, looking for work. He was the descendant of a family of French Huguenots called Hyenveux, who had come to America and changed their surname to the more easily memorable LaFrance.

Truckson LaFrance managed to get a job in the Elmira Union Iron Works, and it was here that he first became interested in steam engines. In 1871–72 he obtained several patents on improvements in rotary steam fire engine design, and the head of the iron works was encouraged to manufacture a steam fire engine.

Soon LaFrance and John Vischer had a small fire engine business going, and their efforts were apparently successful enough to attract the attention of a group of local wealthy businessmen, who duly put up the capital to buy the venture.

Prominent among the new owners of the LaFrance Manufacturing Company were General Alexander S. Diven and his four sons, all, it seems, leading lights in the local community as well as being a family of lawyers. They appointed Vischer a trustee (the equivalent of a directorship), and made Truckson LaFrance mechanical engineer. The purchase took place on 17 April, 1873: no later than July that year, 10 acres of land

(Previous page) A steam pumper at a fire on Broadway, New York, in 1893

(Left) A famous painting of a London fireman: the steps are actually those of Massey Shaw's house

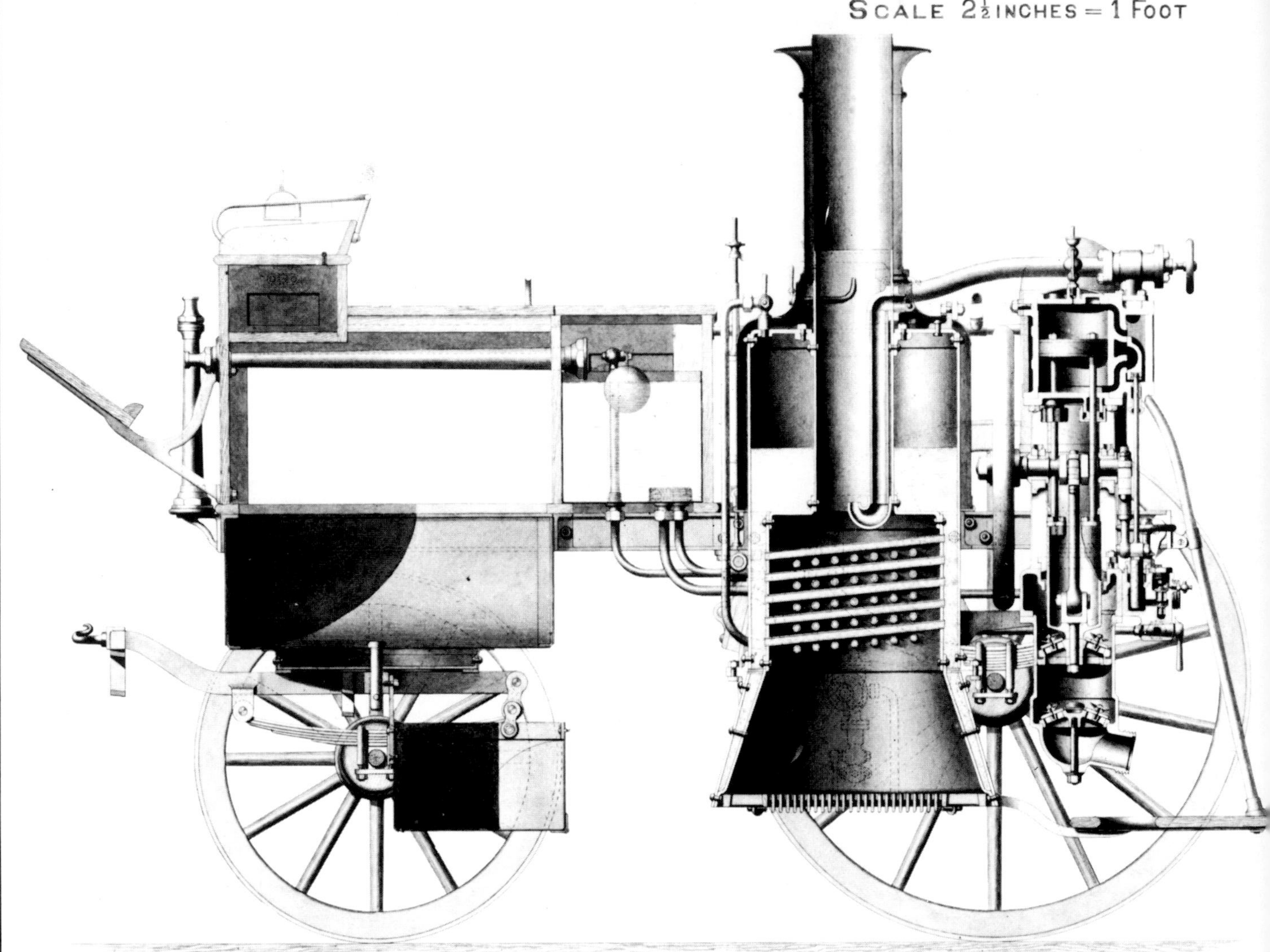

had been acquired and a small fire-engine factory had been completed.

In the early days, the firm turned its hand to any form of engineering that promised to show a profit: apart from 'rotary steam fire engines', LaFrance also made fencing machines, cotton pickers, corn shellers . . . even a railway locomotive that ran on a track mounted on stilts to flatten the contours of uneven ground.

LaFrance was obviously convinced that his fire engines were world-beaters, for in 1878 he sent one to Paris Exposition, hoping to establish an export market. But the design didn't comply with French legislation on boiler plate thickness, and it was forbidden even to demonstrate the engine. The expense of this abortive venture almost broke the LaFrance Company.

They must have been particularly annoyed to read the fulsome reports on a Shand Mason engine which had been demonstrated in Paris a few months earlier by the French agents, Muller & Roux, who,

experimented, in the presence of M le Colonel de Saint Martin and the officers of the fire-brigade of Paris, a very perfect steam fire-engine which raised a pressure of steam of 7 kilos in 7 minutes 35 seconds, and threw a jet of water 33mm in diameter, which reached a height of 52 metres and a horizontal distance of 70 metres . . . on account of its perfect arrangement and excellent construction . . . this machine, which gave such remarkable results, only weighs 1800 kg.

So it *was* possible for a foreigner to comply successfully with the French regulations: but it was a lesson which had proven cripplingly expensive for LaFrance, who had not even enough money on hand to build another engine.

Fortunately, human vanity saved the day. Thomas W. Hotchkiss, a wealthy business-man living in Elmira, was anxious to perpetuate the name of his wife, a noted actress,

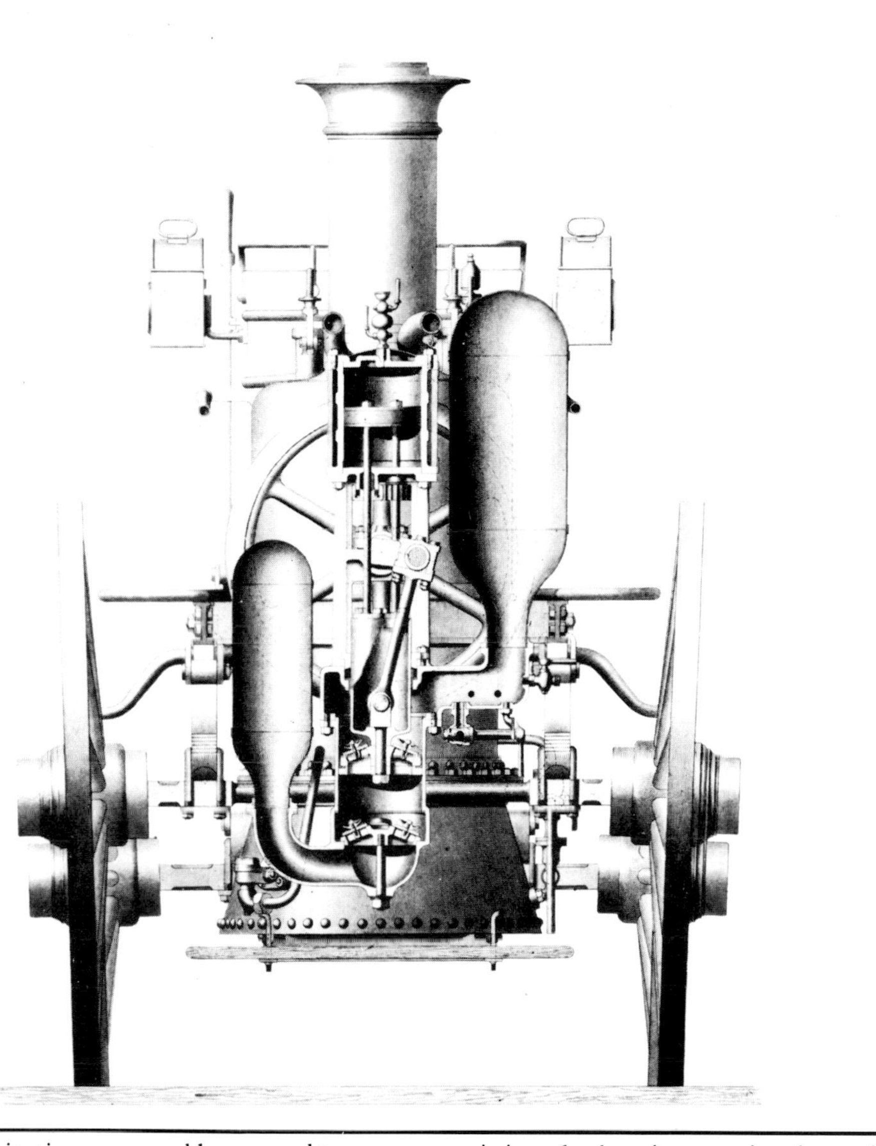

Sectioned views of an English steam pumper of the late 19th century

who also gave recitations . . . and he agreed to finance another engine provided it was christened the *Jeannie Jewell* in her honour.

With the new engine complete, in September 1878, it was decided to achieve some salesworthy publicity by entering it in a three-day fire engine contest at Chicago, where it was shipped in charge of Truckson LaFrance's brother Asa, a travelling salesman for the company (and, it seems, an ace cornettist) and Thomas Hotchkiss. Apparently, the only Jeannie Jewell that went with them was the engine. Perhaps Mr Hotchkiss wanted a rest from his wife's histrionics . . .

The Chicago contest started on 6 September, and the LaFrance company had staked their future on success. By 9 September, the *Jeannie Jewell* was well ahead of the competition. Suddenly, there was a scorching smell, and the engine began to slow: the bearings were overheating, and about to seize. Asa LaFrance acted quickly – grabbing up a spanner, he loosened off the nuts retaining the bearing cap in place, thus easing the end thrust on the shaft and allowing the engine to continue operating. The resultant victory was a contributory factor in setting the company on the road to success; shortly afterwards, Asa won another contest, at Waterloo, Iowa, with this engine, which resulted in the local Red Jacket Hose Company buying the *Jeannie Jewell*.

Two years later, the company became the LaFrance Steam Engine Company: the new title was adopted so that the firm could exploit the new 'rotary nest-tube boiler' invented and patented by Truckson: further inventive progress was made when LaFrance joined forces with Daniel D. Hayes to produce the latter's extension ladder truck, an ingenious device which enabled firemen to climb to new heights while firefighting. Fully extended, the Hayes ladder reached a height of 85 feet: before that, the longest ladders available had an extension of only 50–60 feet. Operation of the Hayes ladder was absurdly simple, for it

This manual engine, The Protector, **was built by the Kelloggs & Cornstock Company, of Hartford, Connecticut**

was raised by turning a long screw passing over a turntable carrying a large nut linked to the base of the ladder. As the screw turned, so the nut moved forward and raised the ladder.

In 1884, the LaFrance Company built its first piston steam engine for the Buffalo, New York, Fire Department, and after this the original rotary-type engine was gradually phased out: soon the new pattern of engine could develop a steam pressure of 80 lb within five minutes of firing up from cold.

Steam engines were certainly becoming both reliable and efficient: in 1878, Eyre Massey Shaw carried out tests in London to determine whether a steamer was cheaper to operate than a manual, and came to the surprising conclusion that it was. Taking a medium-class steam fire engine costing £600, Shaw found that it could discharge 500 gallons per minute at a pressure of 120–130 p.s.i., and cost 2 shillings an hour for fuel and oil. This was equivalent to the work of five manuals, each of which cost £1 10s an hour in operating costs, making a total of £7 10s, which could be swollen to £11 10s when the cost of providing refreshments and 'beer-oh' for 150 volunteer pumpers was taken into account. Thus, at a ten-hour fire, it could cost over £100 to get the same amount of work from manual engines as from a steam engine which cost just £1 to keep running for the same period.

At one fire alone, computed Shaw, the use of steam fire engines instead of the traditional manual machines had saved £617.

Now, instead of tales of destruction, newspaper reports more frequently carried phrases such as: 'The early arrival of the engines enabled the firemen speedily to subdue the flames, and the damage done was not very great.'

With the aid of steam, the most horrifying blazes could usually be brought quickly under control: 'Messrs Marks & Company, dealers in human hair, 232 Essex-road, Islington. Some children playing with lucifers set fire to the front room on the second floor. The children were rescued immediately and the flames extinguished before they could obtain a firm hold on the house.'

Such efficiency was not achieved without training, and firemen were still required to be on duty for numbingly long hours, with few chances for leave. Shaw ruled the London brigade with a rod of iron, and chose his men carefully:

> The ordinary engineer belongs to a trades-union and, even if he wished, would not be allowed to work any hours and all hours, nights and Sundays. He would be of no use at all with his rights as to overtime and his appeals to the central body. The organisation of a fire-brigade must be strictly military, or rather naval, in system. My men know perfectly well that if they are remiss in answering a call, or a 'stop' (a message that an engine is not required), or slow in getting out an engine, the offence will be

visited by reprimand, and will be written against their names in the record-book.

'But,' he concluded proudly, 'many long-service men have never collected such a reprimand.'

Shaw's men had to live at the fire station (though married men were permitted, in the absence of suitable accommodation at the station, to take lodgings nearby). They were on a three-day continuous duty system, with the days occupied in cleaning and maintenance of the fire engines, drill and station work. On the first night, the fireman had to stay in the watch room. Though he was allowed to go to sleep, he had to keep his head close to the electric telegraph which Shaw had instituted as a rapid means of communication between stations. Electric fire alarms, however, didn't appear until the 1880s in Britain, though they were already in use abroad – Montreal, Canada, for instance, had a fairly sophisticated system of public fire alarms in use in the mid-1870s – and, even when alarms were installed, the brigade still relied on messengers arriving breathless at the nearest fire station to gasp out details of a conflagration.

On the second night, the fireman had to man one of the street fire stations which had been taken over in 1867 from the Royal Society for the Protection of Life from Fire, and which housed either an escape ladder (for the obvious step of carrying the escape ladder to the fire on the engine had not yet been taken) or a hose cart. He was allowed to sleep – fully clothed, including his axe – in the watchbox of the station, and was allowed one blanket to make the night more comfortable. Some preferred more exciting comforters, and there are many references in brigade minute books to firemen having been fined 5 shillings for 'entertaining a female in the watchbox of an escape fire station'.

On the third night, a fireman was actually permitted to undress and go to bed, a concession which was withdrawn should the brigade find itself under-strength for any reason.

It was not until 1884 that London firemen could be granted twenty-four hours' leave by their station officer, and forty-eight hours by their district officer.

In 1878, the Metropolitan Fire Brigade moved south of the Thames and took over new headquarters in the Southwark Bridge Road, on a site which had been occupied in the 1700s by a 'grotto' offering entertainments of somewhat dubious merit, and which had then become a workhouse. In 1820, the buildings were converted into a hat factory, and a row of elegant houses added in 1825, one of which Massey Shaw annexed as his private residence when the Fire Brigade moved in (and its ground floor rooms are still preserved today as they were in Shaw's occupancy). The old 'Southwark No 1 Office', little changed itself in external appearance, was still used as a training centre in 1876, and a chilling reminder of its early history can be seen beneath a trapdoor in one of the rest rooms, which conceals the grave of one Pilgrim Warner, who died in the workhouse in 1807.

At the back of Massey Shaw's house was a bay window leading on to a minuscule garden, and it was here each Wednesday that Shaw would invite members of society to watch a special drill display by his men, all clad in the twinkling brass helmets which Shaw had copied from the headwear of the Parisian Sapeurs Pompiers (principal officers had helmets of solid silver).

Even the Royal Family came to watch the brigade's Wednesday displays – the Prince of Wales, later Edward VII, was a particular friend of Shaw's, and a special fireman's uniform was kept for the Prince's use at Chandos Street fire station, near Charing Cross. Shaw would invariably send a brigade vehicle to collect the Prince (who had been taught firefighting at Sandringham by the redoubtable James Compton Merryweather) should a major fire break out, and Edward would dutifully play his part as a volunteer fireman until the blaze was extinguished, when he would hand round Havana cigars to the other brigade members.

Like his Royal friend, Shaw had a private life which was not always above reproach, and it was doubtless with a malevolent twinkle in his eye that W. S. Gilbert made one of his characters in *Iolanthe* sing: 'Oh Captain Shaw! Type of true love kept under! Could

thy brigade with cold cascade quench my great love, I wonder . . .'

But as a fireman, Shaw had no equal: he was vociferous in his dealings with the Board of Works and its Fire Brigade Committee, always attempting to improve the status of his brigade, and to maintain it in the most efficient condition; and he was always willing to face danger with his men, on at least two occasions receiving serious injuries as a result. On 22 June, 1877, he was riding on a steam fire engine which was on its way to a fire in the Isle of Dogs when the horses took fright and bolted. Shaw was thrown from the engine, landing in its path, when its iron-shod wheels ran over his right foot. Fortunately, a few weeks of convalescence at home saw a complete recovery, though an injury at a fire in 1883 left Shaw with a permanent limp.

Outside the area effectively covered by the Metropolitan Fire Brigade, London was defended by volunteer brigades, of which Shaw heartily disapproved: if his men arrived at a fire and found volunteers dealing with it, the volunteers would be summarily elbowed out of the way and told to keep their place. Shaw did, however, agree to the formation of the London Auxiliary Fire Brigade in 1875 – this was a kind of middle-class firefighting enthusiasts' club, with an elected membership and an annual subscription, which acted as an unpaid wing of the official brigade.

Despite Shaw's attitude, many of the outlying volunteer brigades did good work: but they had to rely on voluntary subscriptions for their existence, and this gave sharp characters an opportunity to make ready money. The Press frequently reported cases of money being taken by false pretences to support fictitious brigades – a body calling itself the London and Suburban Fire Brigade, with collectors dressed in Metropolitan Fire Brigade uniforms, existed for thirty years with the sole purpose of parting the public from their money. Bogus collectors even made use of genuine brigade names in their door-to-door fake's progress: in 1876, it was reported, one John Edmunds was sentenced to six months' hard labour for collecting subscriptions to the Notting Hill Fire Brigade, a seemingly laudable activity which netted Mr Edmunds £163 12s 6d until someone remembered that the Notting Hill Brigade had been disbanded for some little while . . .

Possibly Shaw's contempt for the volunteers stemmed from the fact that their crews had not undertaken the rigorous training undergone by his men, training which had altered little when a journalist named Alfred Arkas visited the Southwark headquarters in 1898:

> Sheer physical strength is a desideratum in all branches of public service, but especially so in the Fire Brigade, where lives and property almost always depend on nerve and muscle. Accordingly the strength test (for a prospective fireman) is necessarily a heavy one. A fire escape is brought into the yard, and is rested lengthwise on the flagstones. To a ring-bolt in the stones, a tackle is hooked, the other end being made fast to the foot of the escape. The candidate is then required to haul the escape bodily from the ground into its normal vertical position. It is an immensely trying pull of 240 lb. If the candidate manages it, he becomes a probationer at a salary of 24s a week.
>
> The Southwark instructors reckon that it takes three months' hard work and unceasing drill before a man is competent to leave the yard, even as a fireman of the fourth class. During this period, he is not permitted to attend a fire in any capacity.
>
> No other sort of drill equals in fascination that which the embryo M F B man must go through. Unlike a soldier or sailor, he must undertake many of the actual dangers of warfare on the parade ground. The instruction, conducted by superintendents who have gone through the mill themselves and know every detail of the work, is divided into two parts – theoretical and practical.
>
> The room in which most of the theory is taught is particularly interesting. It contains a half section of every apparatus or device used by or in connection with the brigade. There is a half section of the boiler of the familiar steamer, a half section of a street lamp, indicating the position of a hydrant, and half sections of hose, nozzles,

fire-plugs, flanges, and all the complicated machinery forming part of the various types of engines in use.

That very important part of instruction, the use of steam, is undertaken in the yard, so that practical demonstrations with a steamer under way may accompany the lesson.

Hand in hand with theoretical instruction, a daily grounding goes on in what may be termed emergency drill. To the layman, this is perhaps the most interesting part of the work.

Everything must be rehearsed over and over again. Every movement, every action, must be practised again and again until it becomes automatic, before a man can feel sure of doing the right thing at the right time under circumstances of difficulty and danger. Most of us have seen a fireman descend an escape, bearing on his back a human burden, possibly heavier than himself; we wonder how this is done, but it does not occur to us that this same evolution is practised every day at Southwark in all its separate movements.

Such thorough discipline made the London Brigade the world's finest. But in 1889, the old Board of Works was phased out of existence under the Local Government Act, to be replaced by the newly formed London County Council. Shaw, accustomed to running his Brigade the way *he* wanted, handed in his resignation two years later.

Ceremonial turnout for the Oxford steam pumper, a preserved Merryweather engine

1871:
CHICAGO BURNS

By the 1870s, modern methods of firefighting were coming into use all over the world, and the training methods established in England and America were being put to good effect, thanks in part to the establishment of a firefighting press, represented in Britain by the magazine *The Fireman,* sponsored by Merryweathers, which was to survive for some ninety years.

In its first year of publication, 1876, *The Fireman* carried a report on the state of firefighting in Australia, which seemed, in the larger cities at least to be remarkably up-to-date. Charles Bown, Superintendent of the Sydney Fire Brigade Establishment, commented on the progress of his unusually constituted brigade in 1875:

> During the past year the Northern Insurance Company, who have recently resumed business, have rejoined the Fire Brigade Board of Management, and will now assist in supporting the brigade with the other offices enumerated in this report; a step worthy of imitation by those offices who for years have been doing fire insurance business in this city and whilst enjoying the full benefits and protection afforded by this brigade at all times to contributing and non-contributing offices alike, have hitherto held away from the cost of its maintenance; some on the plea of objection to the charge for entrance fee of 50 guineas, forgetting the fact that in joining they at once participate in the ownership of a plant worth £4000, with a voice in the management of an effective organisation that has now been established over a quarter of a century . . . Last year the members of this brigade were called to attend for one office alone, interested in either stock or building, no less than sixty-four times . . . I have the honour to announce that through the courtesy of Commodore Hoskins, arrangements have been made by which on occasions of disastrous fire in this city, a detachment of men as a fire party from HM ships in port will be dispatched to render aid, as also a detachment of the permanent artillery force from Dawes' Point to assist with the Imperial Steam Fire Engine, stationed at the Naval Depot, George-street North, which for the future will in case of need be available for service at the northern end of the city, the directing and control of which the Commodore has entrusted to the care and discretion of the executive officer of your brigade.

If the Sydney brigade was equipped with the latest type of steam engine, in Japan, northwards across the Pacific Ocean, firefighting methods were decidedly primitive, though as the country had only been open to Western influence since 1853, this is perhaps not so surprising. Around 1860, the Japanese had been using crude wooden squirt pumps which would, in Europe, have looked out of date at the time of the Great Fire of London: but in the late 1870s a visitor to the port of Yokohama found that firefighting technology had made some steps forward, even if the Japanese firemen seemed more than a little foolhardy.

They had a distinctive way, too, of announcing their approach, which was heralded by a shout of 'Hark, here it comes', chanted in time to the footsteps of the firemen. Then the crowd of onlookers would part to allow the firemen to pass: to ensure that they did allow sufficient room, two or three 'night police' would run ahead of the firemen, beating rhythmically on the ground with long iron staves.

Then came the brigade, 'at a good steady trot . . . the officers' silver helmets gleaming in the glare, and the white fire-standards used to rally the firemen where the fire is hottest

(Previous page) A contemporary engraving of the Great Fire of Chicago

(Right) A new York steam pumper of 1911, the Goodwill No. 5

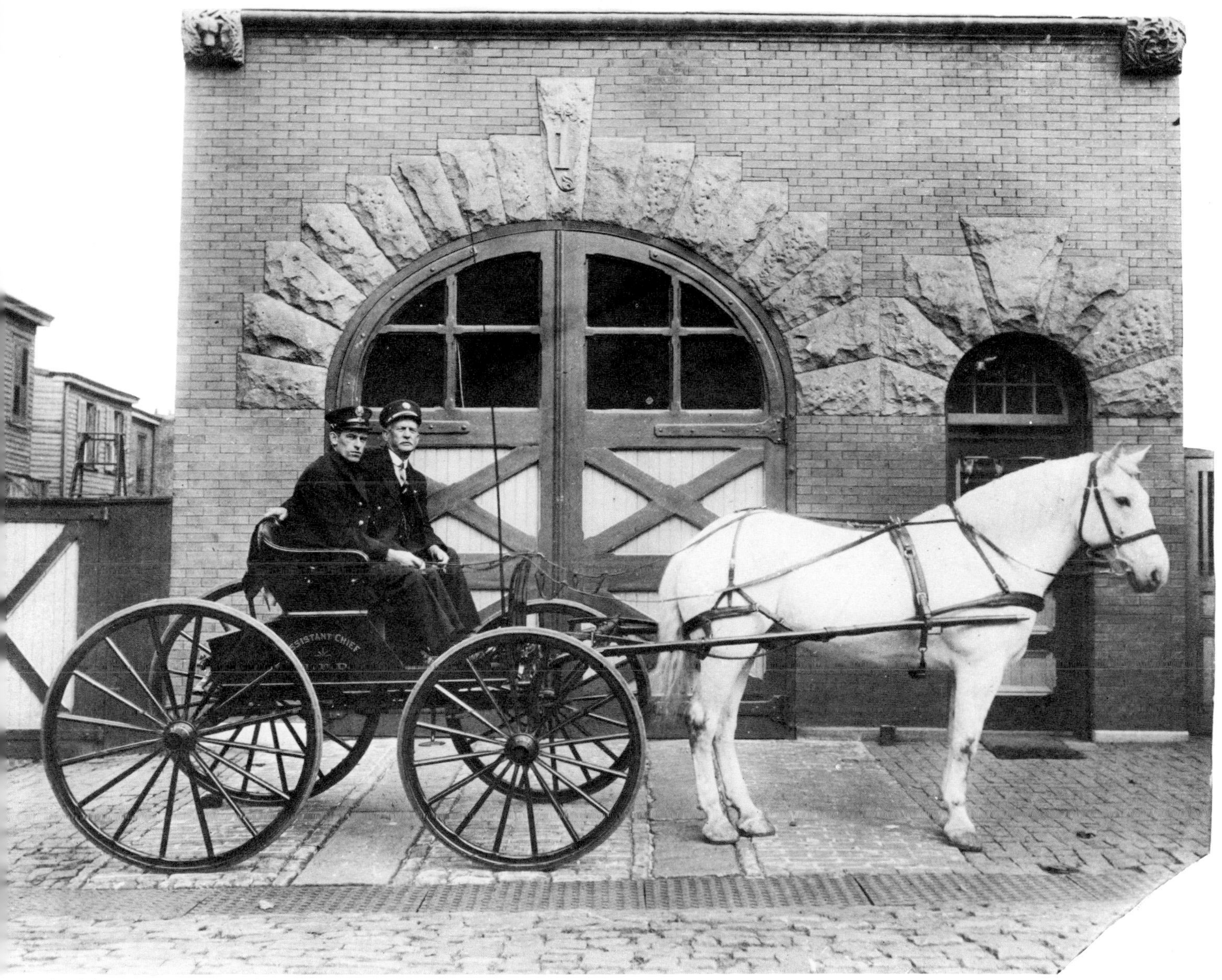

Assistant Chief John A. Stockton of the Camden, New Jersey, Fire Department, operated this elegant buggy in 1913

looking weird and ghostly as they sway and wave backward and forward far above the heads of the approaching body of men'.

Once they had arrived as nearly as possible to the hottest part of the fire, the firemen unslung their engines, which looked like little boxes, from their shoulders and set to work, running up frail bamboo ladders which they leaned against the walls of the houses until the rooftops swarmed with firemen. Only one or two men were required to work each engine, whose output was, to put it mildly, feeble ... but they were useful because they were highly portable, and could be carried on to rooftops, or any other difficult position, and because so many were employed that strength of numbers became a definite factor in their success.

The officers, meanwhile, directed operations from below, 'or from some exalted and very often perilous positions'. But their peril was nothing compared with the men who held the fire-standards, who would stand on the roof of a house far ahead of the main body, with flames leaping and darting around them. As they were the men in the greatest danger, it was reasoned, they were the men who should give the signal to retreat. Until the standard-bearers gave the order to fall back, the other firemen would stay doggedly at their post. The job of fire-standard-bearer was not an enviable one, to judge from eye-witness reports: 'You look up. He is on fire . . . a slight motion of his hand to the firemen below is the only movement he makes. Instantly three or four engines are playing on him, and his burning clothes are extinguished. Until the last chance has gone, he will stay at his post of honour'.

Sometimes, though, the standard-bearers would take nonchalance too far, and a sudden shower of sparks would be their only epitaph.

One wonders whether such an excess of bravery was really necessary when one reads that the Japanese houses were so flimsy that they could be rebuilt within a few hours of

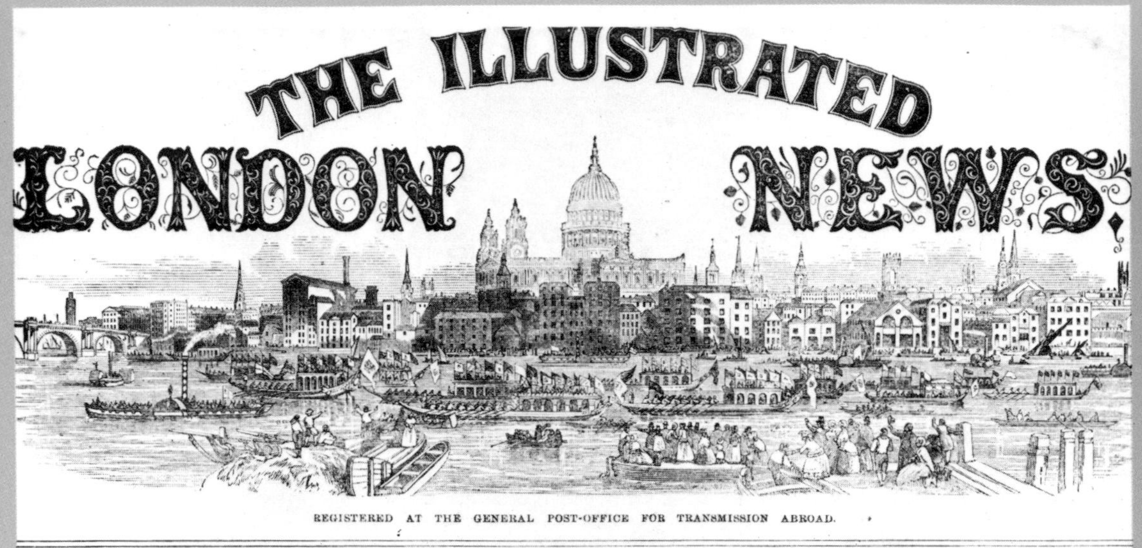

THE ILLUSTRATED LONDON NEWS.

REGISTERED AT THE GENERAL POST-OFFICE FOR TRANSMISSION ABROAD.

No. 1656.—VOL. LVIII. SATURDAY, JUNE 24, 1871. PRICE FIVEPENCE BY POST, 5½D.

Confused fire-fighting methods during the Paris Commune at the time of the Franco-Prussian War: perhaps the man behind the camera is the famous photographer Nadar?

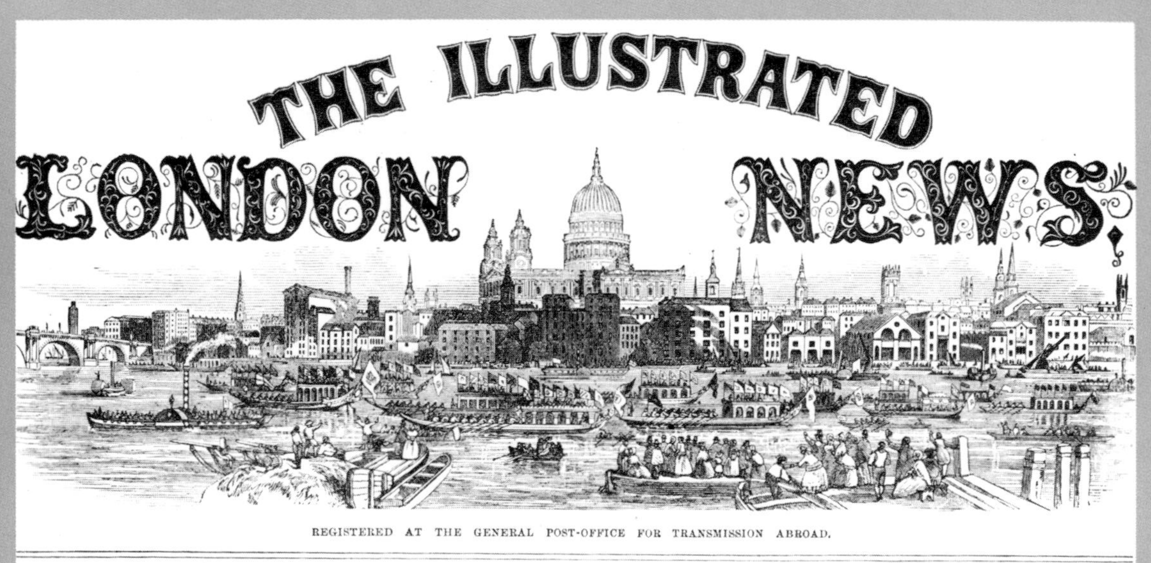
THE ILLUSTRATED LONDON NEWS

REGISTERED AT THE GENERAL POST-OFFICE FOR TRANSMISSION ABROAD.

No. 1678.—VOL. LIX. SATURDAY, NOVEMBER 11, 1871. PRICE FIVEPENCE By Post, 5½d.

One of the main escape routes from the Chicago Fire of 1871 was the Randolph Street Bridge, jammed with people trying to save their possessions from the flames

their being burnt down, a policy which could be taken to excess, as in a major conflagration in Yokohama, nearly an entire street of houses had been replaced after the fire had passed when the wind veered about and caused the flames to retrace their course, completely destroying the newly built dwellings.

Compare this almost childlike attitude to fire prevention to that prevailing in the growing American city of Chicago, which between 1861 and 1871 had built up one of the most modern and best-equipped brigades in the United States. Hand-pumped engines had been largely replaced by steamers, of which the city had acquired a fleet of seventeen. The tale was told of an Irish immigrant, newly arrived in Chicago, who was wakened by the sound of engines on their way to a fire. As they hurtled past, he called to his companion in alarm: 'Wake up! They're moving Hell! Two loads have gone past already!'

In addition to the steamers, Chicago also had four hook-and-ladder carts, 23 hose carts and two hose elevators, platforms which could be raised two stories high so that water could be directed on to upstairs fires. But there were only 185 full-time firemen in the brigade, which had changed from volunteer to professional as early as 1858, one of the first brigades to take this important step. There was also a first-aid cart operated by the newly-formed Insurance Patrol, headed by a young fireman rejoicing in the name of Benjamin Bullwinkle.

And to round off this impressive array of firefighting equipment, the city had just replaced its old hand-crank alarm system with the very latest electric telegraph alarms, backed up by watchmen on the roofs of the fire stations and on the tower of the city courthouse. It would seem that Chicago was well-protected against fire: but the city was really a giant tinderbox just waiting for a spark . . .

The Illinois summer of 1871 had been an abnormally dry one, with only a few showers of rain between July and October to lay the grey dust which settled over the parched city. Chicago had grown rapidly over the past few years, and most of the buildings in the city, if not entirely made of wood, had a high proportion of timber in their construction; roofs were often of timber waterproofed with tarred felt or paper, and even the elevated sidewalks were paved with wood-blocks.

And the city's industries – furniture-making, carriage-building, wood-working and paint manufacture – all used highly inflammable materials.

Warning of impending doom came on 7 October, 1871, when a lumber mill caught fire, and the resulting blaze almost entirely wiped out four city blocks. This fire was barely under control when a tiny, unobserved incident set off a far more serious conflagra-

tion. According to popular legend, a cow kicked over an oil lamp in the barn behind a cottage on De Koven Street, in the working-class West Side of the city from which Mrs Catherine O'Leary ran a little dairy business. Whatever the actual cause of the fire, within a few minutes it had spread from the barn to several neighbouring houses, fanned by a strong breeze.

First firefighters on the scene were the crew of the hose-cart *America,* whose foreman had spotted the blaze soon after it had started, followed shortly by the steamer *Little Giant,* whose lookout had also seen the flames in the night sky. *America,* working at mains pressure, was only partially effective against the fire; the main lookout station on the roof of the courthouse miscalculated the position of the fire and sent out a wrong alarm which caused unnecessary delays in steamers arriving on the scene. Furthermore, the brigade had not had sufficient time for men or machines to recover from the previous night's

(Left) A floral welcome to delegates at the Union Fire Company's Centenary Convention in 1889

blaze, and though the fire was soon ringed by the steamers *Waubansia*, *Illinois*, *Chicago*, *Economy* and *Little Giant*, aided by the hose carts *America*, *Tempest* and *Washington*, and the hook-and-ladder wagon *Protection*, the flames were gaining an uncomfortably secure hold. *America* was connected up to a fire-plug which was usurped by *Waubansia*, as fire brigade etiquette required mains-pressure hose carts to give way to the more powerful steamers, a sensible enough notion which unfortunately broke a weak but nonetheless vital link in the chain of defence, which was simultaneously ruptured when *Chicago's* pump seized up (though a sharp thump with a hammer subsequently freed it) – and through these two gaps the wind sent the flames and sparks, causing houses across the street to catch fire. And the fire hose was old and worn, and liable to burst under pressure . . .

An hour later, two blocks were on fire, and the wind was hurling blazing embers across the sky: as the flames spread, the engines moved into independent groups each working to try and keep the fire in check.

But already they were beginning to be too thinly spread: a huddled community of shacks, pigsties and sheds a couple of blocks north of O'Leary's barn was now ablaze, and though two more steamers, *Frank Sherman* and *Long John* had now arrived on the scene, the firemen could only hold back the fire on the south and west sides. To the north and east, fanned by a wind which was now gusting up to 60 m.p.h., it was already getting out of hand. It was little more than an hour since the O'Leary barn had caught fire . . .

Several of the steamers attempted to form a new barrier to the north of the flames, but were beaten back by the heat. The *Waubansia* had to be dragged away by hand, for there was no time to bring up its horses, which were tethered at a safe distance. As the fire gnawed its way north-east, the wind grew in violence, and drove the sparks into an area of lumberyards and furniture factories. The owner of the yards, an ex-fire marshal, attempted to flood them to form a firebreak, but to no avail. Now the engines were attempting to throw their jets into the face of a gale against which it was difficult to stand upright: but the men stood firm until they were forced to retreat.

The crew of the *Frank Sherman,* for example, held on until their hose began to smoulder, and the flames were singeing their hair, and the legs of their horses, which they had prudently kept harnessed. But in three hours the fire had gained seven city blocks, and had jumped the river which the firemen had hoped would form a natural firebreak, spreading into the South Side, which contained the principal residential and business quarters of the city.

One of the steamers which had been trying to halt the progress of the flames, the *Fred Gund,* was coupled to a hydrant near two liquor stores, whose stock-in-trade was burning with a fierceness that seemed to turn the water from the hose into steam before it hit the ground. The crew, aided by volunteers, began to try to drag the engine away from the flames, but could not uncouple the hose because of the heat. Then they found that the wheels of the engine were scotched by lumps of coal: just as they removed these, a sudden eddy of flame sent them scuttling down the street. They tried to get back to the engine, but as they struggled towards it, the hose suddenly snapped, burned through at the hydrant, and the firemen could only watch helplessly as the *Fred Gund* rolled ponderously into a blazing sidewalk and was then buried in a redhot avalanche of bricks as the side of a building collapsed . . .

Once across the river, the fire redoubled in intensity, and by 1.45 a.m. the 'fireproof' courthouse was ablaze: at 2.15 it collapsed in ruins. Refugees fled from the threatened area as buildings were gobbled up like 'the cardboard playthings of a child': one curious sight was an Indian file of boys, each carrying a coffin on his head, followed by an undertaker driving a waggon on which was a large casket. Many moved their possessions away from the flames, only to have them destroyed later as the fire continued its inexorable course, crossing the South Branch of the river which, it had been devoutly hoped, would have formed a second barrier to the flames.

The steamers found it impossible to play their jets against the wind, which was cutting off the stream only a few feet from the nozzle: then, as the steamers *Economy* and *Rice* fought to save the railway marshalling yards, their hoses went limp. The waterworks, a monstrous folly in debased Gothic which drew its supplies through a tunnel into the heart of Lake Michigan, had caught fire. Once the steamers had pumped the mains dry, that was that. Some engines managed to form relays from the lake shore, and Chief Fire Marshal Williams had ordered some of the latest firefighting appliances, four-wheeled chemical engines, which could expel water under pressure from carbon dioxide produced by a reaction inside the engine's reservoir, into action as standbys along the river. He had even telegraphed to Milwaukee to send steamers and firemen as reinforcements for the hard-pressed Chicagoans. But the firemen could now only carry on a limited holding action, keeping the fire in check only where they could dip their leaky hoses into the lake.

It was a change in the weather which eventually caused the end of the Great Fire of Chicago. A little over 24 hours after the holocaust had started, a fine drizzle began to fall, damping down the flames: the following morning, the citizens of Chicago began to count the cost. An area four miles long and a mile wide had been razed to the ground, 100,000 people had been made homeless, and

17,500 buildings had been destroyed (but the O'Leary cottage still stood, only slightly singed . . .); perhaps two or three hundred people had died.

Among the offers of help which poured in from all over America was the loan of two of the very latest pattern steam fire engines, one made by the Amoskeag Manufacturing Company of Manchester, New Hampshire, the other by the Silsby Manufacturing Company of Seneca Falls. These guarded the city for several weeks while the Chicago engines were restored to full working fettle.

The Chicago Fire Department, it seems, was determined not to let the disaster repeat itself, and within a few years, the brigade had doubled in strength. In 1878, there were 386 men and officers, controlling 33 steam fire engines, 151 horses, 1850 feet of ladders, 35,425 feet of hose and 466 alarm boxes. There were also two fire patrol companies, each with two waggons and 20 men, 'properly officered and having all the effective appliances for extinguishing fire and protecting personal property from fire'. Additionally, that year Fire Marshal Benner inaugurated Chicago's first school for firemen.

Perhaps all these precautions had been taken because of man's inability to profit from experience: in the years since 1871, much of the burned area had been rebuilt . . . with wooden houses.

In fact the Chicago Fire Department had reason to be proud of their resources. Compare their state of readiness with that of Dijon, in France, a town of some 40,000 inhabitants in 1878: its sole fire defence consisted of half-a-dozen tub pumps of positively medieval design. Each of these two-wheeled museum pieces held only forty gallons of water. Enough, when working full out, to last for precisely two minutes . . .

(Left) Chemical engine of the Laurel No. 1 Volunteer Fire Department, of York, Pennsylvania, 1911

(Below) Hook and ladder truck of the York (Pa.) Fire Department, also photographed in 1911

HEYDAY OF THE HORSE-DRAWN STEAMER

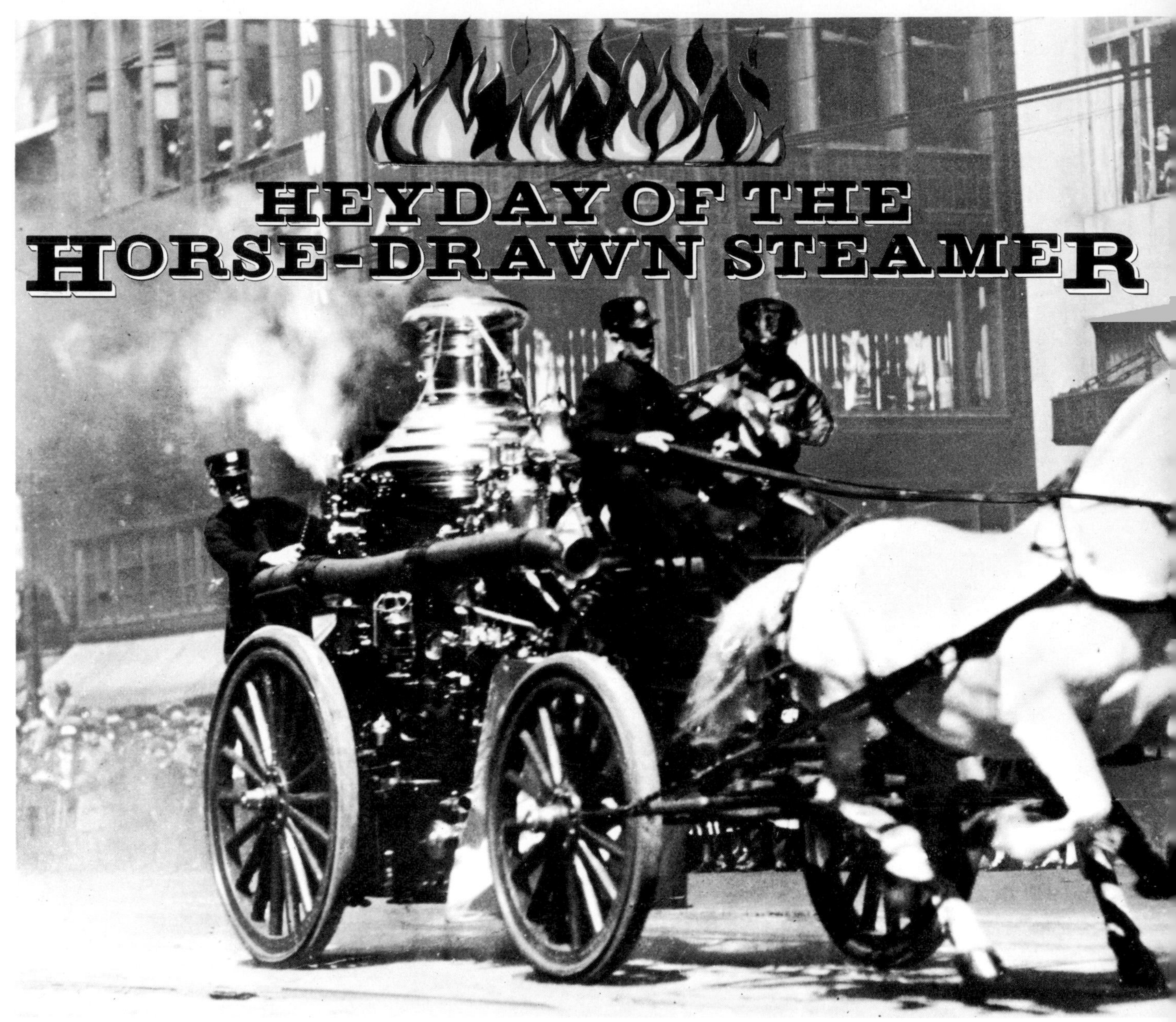

Spectacular turn-of-the-century American steam pumper – probably a La France – with three horses harnessed abreast

The era of the horse-drawn steam fire engine lasted forty years: during that period, with no serious challengers to its supremacy, the horse-drawn steamer changed little in overall design, though detailed engineering refinements were, of course, made from time to time. The standard British pattern of steamer, built by both Merryweather and Shand Mason, had a vertical twin-cylinder engine at the rear. The type had its origins in the single-cylindered London Brigade Vertical engine built by Shand Mason in 1863; Merryweather remained faithful to horizontal cylinders until 1885, when the twin-vertical-cylinder *Greenwich* model was produced, named after the new factory which Merryweather's had opened in 1876 to cope with the increase in demand for their products (the original Long Acre works, condemned as unsafe by the Board of Works, had been rebuilt in 1873, while the Lambeth factory, in one of the most unsavoury streets in that borough, closed in 1879). The *Greenwich* 'Patent Double Cylinder High Speed Steam Fire Engine' could deliver 750 gallons of water a minute and was, according to its manufacturers 'eagerly sought after in South American and Russian markets'.

Even more popular was the *Greenwich Gem,* introduced in 1896 at the London Fire Tournament, which was built in sizes from 200 to 500 gallons per minute pumping capacity: in 1896, too, Merryweather announced their new *Hatfield* reciprocating pump, which had three cylinders mounted at 120 degrees inside a hexagonal casing, taking water from a common suction chamber, and delivering it *via* a common passage.

Rubber disc inlet and outlet valves made the *Hatfield* pump remarkably uncritical in its diet, a major advantage when pumping dirty water from ponds or ditches. In Hong Kong in 1927, a Hatfield ran continuously for a fortnight supplying water to striken areas after a violent storm had devastated the colony. One Hatfield pump was returned for

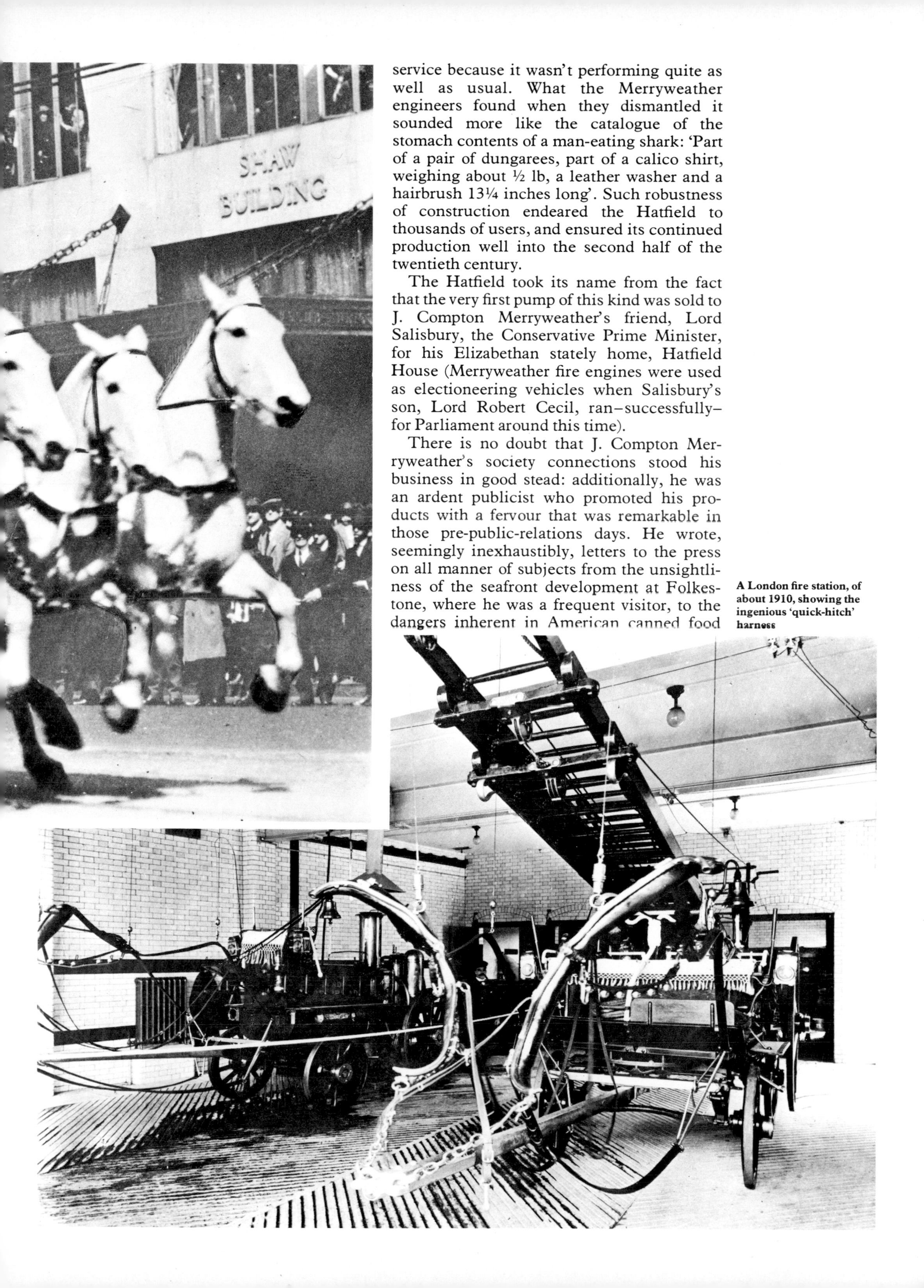

service because it wasn't performing quite as well as usual. What the Merryweather engineers found when they dismantled it sounded more like the catalogue of the stomach contents of a man-eating shark: 'Part of a pair of dungarees, part of a calico shirt, weighing about ½ lb, a leather washer and a hairbrush 13¼ inches long'. Such robustness of construction endeared the Hatfield to thousands of users, and ensured its continued production well into the second half of the twentieth century.

The Hatfield took its name from the fact that the very first pump of this kind was sold to J. Compton Merryweather's friend, Lord Salisbury, the Conservative Prime Minister, for his Elizabethan stately home, Hatfield House (Merryweather fire engines were used as electioneering vehicles when Salisbury's son, Lord Robert Cecil, ran–successfully–for Parliament around this time).

There is no doubt that J. Compton Merryweather's society connections stood his business in good stead: additionally, he was an ardent publicist who promoted his products with a fervour that was remarkable in those pre-public-relations days. He wrote, seemingly inexhaustibly, letters to the press on all manner of subjects from the unsightliness of the seafront development at Folkestone, where he was a frequent visitor, to the dangers inherent in American canned food

A London fire station, of about 1910, showing the ingenious 'quick-hitch' harness

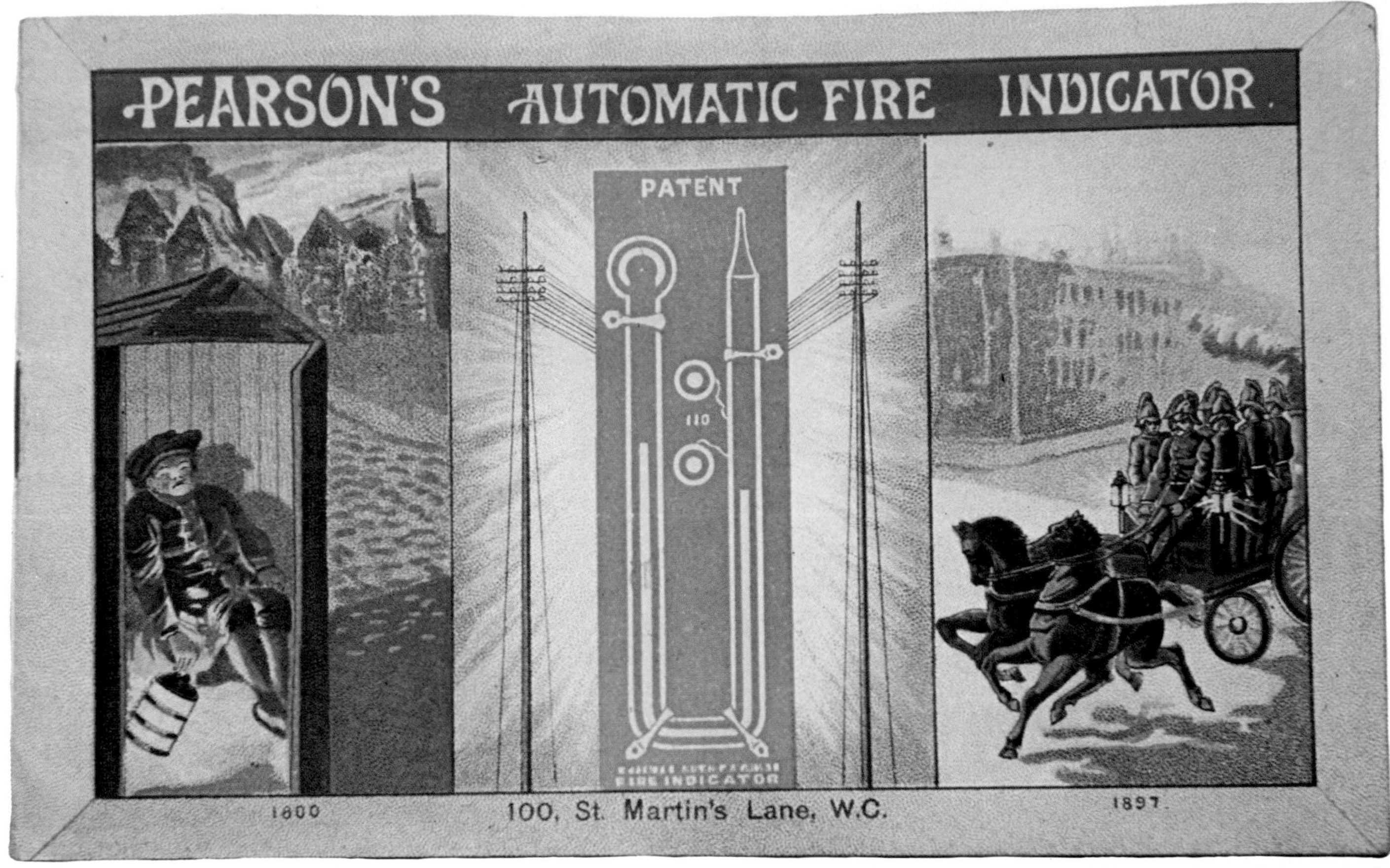

Cover of a fire alarm catalogue of 1897

(Opposite, top and centre) Edwardian firemen took justifiable pride in the smartness of their engines – even their helmets were of polished brass!

(Opposite, below) Early fire alarm switchboard at the headquarters of the London Fire Brigade. Unfortunately, passing trams were liable to trigger off false alarms by vibrating the indicator flaps shut . . .

('American canners are using poisonous tin plate'). Anything, in short, to get the name of Merryweather into print. He travelled the world learning about firefighting methods and requirements in other countries, building up a thriving export business. On one of his trips he presented the Khedive of Egypt with a Merryweather Patent Travelling Fire Escape, a device of the kind beloved of Victorian globetrotters which, while ostensibly a large Gladstone bag, in emergency could be converted into a sort of breeches buoy, in which the traveller could slide down a cable from his burning hotel room. In return, the Khedive made Merryweather a Commander of the Imperial Order of Medjidish (Third Class): the impressive certificate recording the award is still preserved in the Merryweather archives.

Another example of Merryweather's interest in overseas markets was the supplying of steam fire engines to the city of Paris during the Franco-Prussian War. The waterworks were behind the German lines, so the steamers pumped water from the Seine for the use of the citizens.

He was convinced that language differences were one of the great barriers to international commerce, and so was an advocate of that early *lingua franca,* 'Volapük'. He wrote a letter on the subject to the *Westminster Gazette,* and to show his enthusiasm, transcribed his signature into Volapük as 'J. C. Yofikstom'!

It was no coincidence that the period during which 'J. C. M.' was at the head of the company saw the name of Merryweather predominant in the manufacture of firefighting equipment, for James was an engineer by training. He had started as an apprentice to an engineer and brass finisher, and had later spent some time in the Wolverton works of the London & North Western Railway.

When Merryweather shares came on to the market, they were eagerly snapped up by members of the peerage, and the firm's products attained such a cachet from their association with the pillars of society that when the son of Major Joicey of Sunningdale Park was asked what he would like as a present on his twenty-first birthday in 1894, he replied: 'A Merryweather fire engine'.

James Compton Merryweather's recreations were those of most Victorian men of property: he owned two racehorses called, symbolically, 'Fireworks' and 'Running Fire', but both turned out damp squibs.

He kept copious cuttings albums, pasting in the items with his own hands, preserving everything from news of his latest products to invitations to events as diverse as Queen Victoria's Diamond Jubilee and the local cycle club's day outing. Every new Merryweather machine was photographed for inclusion in massive albums, and in many of the pictures old James can be seen, leaning against the engine with a proprietorial pride, his low-crowned billycock hat in his hand.

HASLEMERE.

J. C. Merryweather himself, testing a lightweight steam pumper destined for Rome

His activities earned him the well-merited nickname of 'The Fire King', and he was still working in the factory a week before his death in 1917 at the age of 77. His workmen mourned him as a just and good employer, who would often wander through his factory, addressing them all by their first names and genuinely interested in their welfare. Merryweather left £2000 to be divided among his workmen, plus £500 for the women and girls, 'as a small recognition of the patriotism of Englishwomen during the war'. There was no son to succeed him: his place was taken by John Henry Osborne, the overseas representative of the company.

There was certainly drama inherent in the sight of a steam fire engine, drawn by a team of straining horses, dashing through the streets, its crew shouting 'Hi-Ya-Hi' to clear the way, and smoke and sparks streaming from the funnel. Some engineers, apparently, were not above throwing a handful of sawdust on the fire to create more spectacular sparks . . .

And it was this drama which made the fire engine a natural 'star' for the latest entertainment sensation of the 1890s, the bioscope. One Edison cameraman was, it is said, so intent on filming a speeding fire engine that he failed to lift his camera out of its way quickly enough, and the apparatus was smashed to flinders by the engine. But the precious film magazine miraculously survived intact, and the film proved so exciting that the front row of the audience would rush for the exit, convinced that they were about to be run over!

On a slightly higher plane was a later Edison 'motion picture attraction', *Life of an American Fireman,* produced by Edwin S. Porter, which dramatically showed the rescue of an 'imperilled woman and child' from a blazing building. Included in this epic was film of the latest idea in firefighting, a metal pole passing through the floor of the firemen's sleeping quarters, down which they could slide to the engine house when the alarm was sounded.

The film showed an engine turning out in 'the almost unbelievable time of five seconds': but the film was not so very far from the truth here, as, in addition to the special quick-hitch harness, the fire horses had an almost uncanny instinct in sensing emergency. Yet both in Britain and America, fire companies hired their horses from jobmasters rather than owning them: in London, the famous Thomas Tilling supplied the Fire Brigade horses, usually big greys, while in Edinburgh it was the stables of John Croall which provided motive power for the engines. When an alarm rang in the station, it also rang in Croall's stables, and fresh horses were brought up as standbys. When Edinburgh obtained its own horses, they were trained to go to their places in front of the engine as soon as the alarm sounded: and in New York's stables, it is recorded, there was one old horse who was so sagacious that he would dash to

May 1, 1895]

THE FIREMAN.

223

FIRE BRIGADE EQUIPMENT.

Messrs. MERRYWEATHER make a Speciality of Firemen's Uniforms and Equipments.

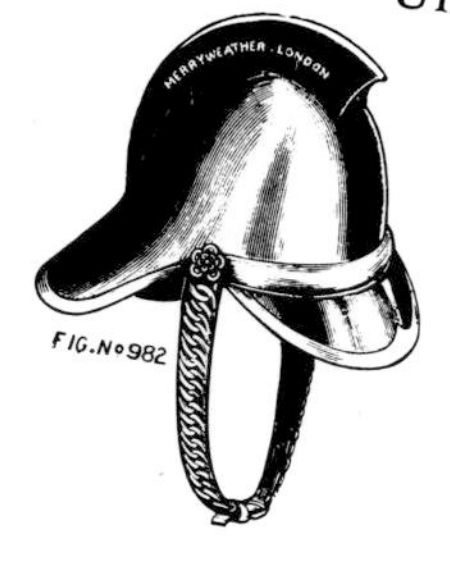

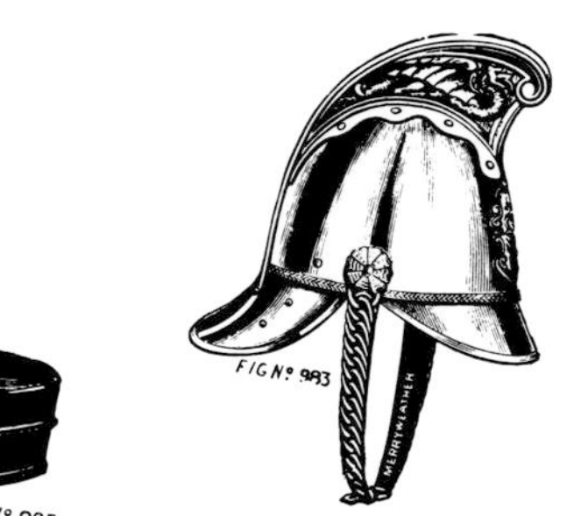

FIG No 985

UNIFORMS OF ALL DESCRIPTIONS FOR OFFICERS AND FIREMEN.

Helmets of All Designs and Shapes for Home and Abroad, in Leather, Brass, Cork, Felt and Plated, for Firemen and Officers, and of Specially High Finish for Presentation.

Undress Caps of All Patterns for Officers and Men. **Rubber and Leather Boo...**

OVERALLS, AXES, BELTS AND POUCHES.

Epaulettes, Ornaments, Medals, Whistles, Speaking Tru...

BUTTONS, ROSETTE...

Fancy Buck...

Firemen from the Pro...

where they can inspect

Write to GREEN...

FIRE BRIGADE EQUIPMENT.

Messrs. MERRYWEATHER make a Speciality of all kinds of Fire Brigade Uniform and Equipment.

Helmets of all Designs and Shapes for home and abroad, in Leather, Brass, and Plated, for Firemen and Officers, and of specially high finish for presentation.

Undress Caps of all Patterns for Officers and Men.

Rubber & Leather Boots.

Axes for Firemen and Officers, and of specially high finish, beautifully Embossed and Plated for presentation.

Belts and Pouches.

Epaulettes.

Ornaments.

Medals.

Whistles.

MERRYWEATHER. FIG. No 1251

Firemen from the Provinces visiting London are cordially invited to call at their Show Rooms

63, LONG ACRE, W.C.,

where they can inspect Samples of all the Latest Novelties in this Department.

Write to GREENWICH ROAD, S.E., or 63, LONG ACRE, W.C., for full Special Descriptive Lists.

Merryweathers not only built fire engines — they also supplied 'all kinds of Fire Brigade Uniform and Equipment'

Country houses were often self-sufficient in fire-fighting terms: some wealthy landowners, like the Duke of Portland, even operated their own private fire-brigades

his engine as soon as anyone came into the station wearing a fire-helmet!

Reliance on horse traction was fine for the larger brigades, but the smaller volunteer forces often found the acquisition of horses a problem in an emergency, for they could not afford to keep horses 'on the permanent strength', and it was these smaller brigades who would lead the way in the rapid discarding of the horse-drawn fire engine in favour of new methods of traction.

But before its days were ended, the horse-drawn steamer assisted at many spectacular fires. And perhaps the most spectacular of all was the great Cripplegate fire of 1897, said to be the worst fire in the City of London since the Great Fire of 1666. It started on the first floor of a warehouse occupied by a firm of gas mantle makers on the corner of Hamsell Street and Jewin Street, in the middle of an area of warehouses piled high with goods such as dress fabrics, silks, furs, feathers and Christmas novelties; the cause of the outbreak was variously given as a gas explosion or arson. Whatever the origins of the fire, it soon gained a strong hold on the building, aided by the fact that the locals were somewhat tardy in giving the alarm.

By the time the first engines arrived on the scene, the fire was well alight: then, according to eye witness reports, more problems arose . . .

At first, the requisite pressure of water could not be obtained, while they were also severely handicapped in not being able to secure a suitable coign of vantage from which to play upon the flames. Building after building was rapidly reduced to a mass of smouldering dust and ashes. Steamer after steamer drove up, till within a very short space of time there were no less than sixty engines with three hundred firemen on the spot, drawn from every available station in the metropolis, pumping hundreds of tons of water upon the burning buildings. Yet, notwithstanding these

Colourful turn-of-the-century catalogues issued by fire engine manufacturers

determined efforts, the fire continued with unabated fury, and nothing could be done to arrest its progress.

Hamsell Street was soon nothing else but a heterogeneous pile of masonry, twisted ironwork, and half-consumed wares, and the fire, having burned through the back premises on to Well Street, soon laid waste the stately buildings lining each side of that thoroughfare. It then advanced towards Jewin Street, and thence to Jewin Crescent. It was here that the firemen for the first time were able to gain the upper hand, and once having broken the backbone of the conflagration, they soon made rapid headway in completing its entire subjugation. What a terrible scene this desolated district presented! It appeared as though some invading army had passed over it.

For purposes of subjugating this fire 15,000,000 gallons of water, approximating 67,000 tons in weight, were consumed, all drawn exclusively from the mains of the New River Company, for which they received no payment whatever.

The losses of the fire were enormous. Apart from the loss of revenue to the water company, the fire destroyed nearly a hundred buildings over an area of 100,000 square feet, causing well over a million pounds' worth of damage. It was a spectacular climax to the age of the horse-drawn fire engine.

The press, however, was critical: 'London, in the matter of skilful dealing with gigantic conflagrations, is sadly behind the times. Our firemen are well-drilled, active and daring, but the means at their disposal for grappling with the demon of fire are utterly inadequate. Other great cities of the world have long ago adopted chemical engines and water-towers, and the drastic but sure method of isolating a fire by knocking down the surrounding buildings. The old notion, to which our brigade is bound, is – save the building and get the fire out: the new and better principle is – get the fire out first, and any how, but get it out.'

THE SELF-PROPELLED ENGINE

(Above) Claimed to be the world's largest fire-engine, this Amoskeag operated in Hartford, Connecticut, in the 1890s

(Opposite page) A trio of Merryweather Fire Kings, including an 1899 example destined for Mauritius (lower photo). The Broadrick engine, built for the Singapore Fire Department, is fitted with wire-spoked wheels to withstand termite attacks.

Apart from Reuben Plass's abortive attempt in America in the 1870s, the newly developed internal combustion engine was first applied to a fire engine in 1888. Gottlieb Daimler, who had patented the first high-speed petrol engine in 1883, and fitted such a unit to a carriage in 1886, was friendly with Wilhelm Kurtz, proprietor of an old-established bell-foundry and fire engine works in Stuttgart. Kurtz, indeed, had provided the castings for Daimler's first experimental power unit, and suggested that the internal combustion engine was ideally suited to driving a fire pump, especially in Germany where most of the brigades were run on a voluntary basis.

Daimler, who saw his invention as a universal driving force with a wide variety of applications, lost little time in putting his friend's suggestion into practice, and in July 1889 patented the concept: that summer, the Daimler fire engine (which used its 4 h.p. engine just to drive the pumps, still relying on horses for traction) was demonstrated at the German fire brigade display at Hanover. It created a considerable impression among the experts who were present, for it could be brought into action with far greater speed than a steamer, and needed only two men to operate it. Weighing 3142 lb against the 6614 lb of a steamer of comparable power, the 1892 6 h.p. Daimler fire engine could pump around 100 gallons per minute to a height of 98 feet, and was so reliable, despite its complex hot-tube ignition, that one of these units was still in regular service in 1925.

Daimler petrol fire engines were soon being exported: in 1891 the company's American agent, William Steinway (perhaps better known for his pianos) announced that his factory on Long Island could supply Daimler 'gas and petroleum motors for street railroad cars, pleasure boats, carriages, quadricycles and fire engines', while at the 1893 Columbian Exposition in Chicago a 6 h.p. fire engine was one of Steinway's exhibits. However, Steinway's death in 1896 brought an end to this ambitious venture.

In 1895 Sir David Salomons organized Britain's first motor show, at Tunbridge Wells in Kent: here, too, a Daimler fire engine was a star attraction. Daimler fire engines began to appear in France in the early 1890s, as well. The use of petrol engines to drive the pumps on horse-drawn vehicles was not confined to Daimler: Merryweather built a few such units in the early 1900s, as did the Hungarian firm of Teudloff Dittrich, founded in 1885 as a manufacturer of manual fire engines.

Probably the first successful fire engine to use a petrol engine for propulsion as well as pumping was shown at the 1898 Paris Salon de l'Automobile. Built by Cambier et Cie, of St-Maurice, Lille, France, on the 'Systeme du Capitaine L. Porteu et Th. Cambier', it had a four-cylinder engine driving through a two-speed and reverse gearbox with twin chains to the rear wheels. It could attain 8 k.p.h. in bottom gear, 15 k.p.h. in top: its double-throw fire-pump was driven by bevels from the crankshaft.

It was the first of many European self-propellors to appear in the last years of the nineteenth century, although America, with a long tradition of steam-powered self-propellors, had already produced some spectacular machines. In 1894 the Hartford, Connecticut, fire department possessed what was confidently claimed to be the largest steam-powered fire engine in the world. Built by the Amoskeag Corporation, of the Manchester Locomotive Works, Manchester, New Hampshire – who had been building self-propellors since 1876 – this behemoth could throw a jet of water almost 350 feet, and was linked to a boiler in the basement of the engine house when not in use, so that its water was always hot. Thus, within two minutes of firing up, the Amoskeag could generate enough steam to drive it at 31 m.p.h.

It was steam power which interested James Compton Merryweather, too: his Greenwich Gem engines needed only steering mechanism and some form of driving gear to convert them to self-propulsion, and during 1898–9, Merryweather obviously had some such scheme in mind, spurred possibly by correspondence such as that he received one day in 1899 from Charles T. Crowden, a monumentally unsuccessful motor-car maker from Leamington Spa, who sent a cutting from the local paper showing a cartoon enigmatically titled 'The fire-engine converted into a motor car'.

'Dear Mr Merryweather, Is this anything in your line?' hopefully queried Charles T., affixing a signature in violent purple ink, but it seems as though Merryweather had already taken the decision to build a self-propelled engine, and didn't need any assistance from Mr Crowden. The latter, however, did convert a Merryweather Gem to self-propulsion soon

60

Gola & Conelli

INGEGNERE

GOLA & CONELLI

MILANO - VIA DANTE N.16

APRILE 1899

MAGGIO

GIUGNO

POMPE D' INCENDIO A VAPORE ED A MANO, TUBI ED ATTREZZI PER POMPIERI

della Casa

"MERRYWEATHER & SONS L.td „ di Londra

FONDATA DA OLTRE 200 ANNI - 66 MEDAGLIE D'ORO E DIPLOMI D'ONORE

Pompe d'incendio a vapore fornite in Italia ai seguenti:

MUNICIPII di Venezia, Firenze, Prato Toscana

ARSENALI di Napoli, Spezia, Castellamare e Venezia.

CANTIERI di Napoli, Spezia

DIREZIONE ARTIGLIERIA e TORPEDINI di Napoli

CORAZZATA "Italia „

Sigg. I. SCOTT & SON - Napoli.

Another view of a 1904 Fire King on test: this is the Norwich Union Insurance Company's Worcester engine

(Opposite page) Horses were still to the fore when Merryweather's Italian agents issued this calendar in 1899: but within three years, self-propelled steam engines dominated the company's output

after, for the local branch of the Norwich Union Insurance Company.

Indeed, the first sketches of the proposed Merryweather self-propelled steamer had appeared in May 1898, and in the interim the workshops of the North Eastern Railway Company at Hull had converted an 800 g.p.m. Merryweather steamer to self-propulsion.

There was obviously a growing need for a reliable powered fire engine. Captain Lionel de Latour Wells, who now commanded the Metropolitan Fire Brigade told the *Sunday Daily Mail* in 1899: 'If only we could procure an absolutely reliable motor, we could do away with horses tomorrow . . . Its most immediate probable use is for suburban brigades, where the traction of fire engines is largely dependent on the use of gentlemen's horses'.

The Parisian Sapeurs-Pompiers were not so hesitant. On 1 July, 1899, the *Morning Leader* wrote:

> The Paris Fire Brigade now uses a motor fire engine. This is no mere adaptation of the ordinary motor, but is an absolutely unique auto-motor fire engine, highly complex in construction and invented by Capitaine Cordier, the engineer-captain of the Brigade. It is a low vehicle, can enter the narrowest streets, and on being brought to a standstill, the same force which has driven the car can be instantly utilised for working the pumps. Paris firemen claim for it an advantage over the horse engines of two or three minutes starting, and as much more in getting to work.

The ingenious Capitaine Cordier, apparently, also constructed a number of *fourgons éléctriques* for the Sapeurs-Pompiers.

In September 1899, the first self-propelled Merryweather steamer was complete: it could send a jet of water 150 feet high at the rate of 300 gallons per minute, and was capable of generating a working head of steam within six minutes of starting from cold. On test the following month, it proved capable of ascending Blackheath Hill at a steady 10 m.p.h., and could attain a breakneck 20 m.p.h. on the level. It was destined for use in India, and the automotive press bemoaned the fact that such a machine was not in service with a British fire brigade.

The eclipse of the horse-drawn steamer was rapid: by 1902 Merryweather were concentrating on the self-propelled *Fire King* steamer to such an extent that the horse-drawn unit was almost extinct: that year they delivered the London Fire Brigade's very last horse-drawn steamer, a specially designed lightweight unit for use in hilly districts. Fifteen years later, the last London steamer attended its last fire, at Peckham. In rural districts, of course, the horse-drawn units lingered for a few years more, but by the end of the 1920s almost all were extinct. Probably the last horse-drawn engine in service was an ancient manual owned by the Bishops Castle, Shropshire, brigade, still used in 1931,

A·2·KJ

though other impecunious brigades attempted to update their steamers by pulling them behind motor vehicles. One Suffolk brigade even removed the wheels and axles from its ancient steamer and fastened the cumbersome bulk on to the rear of a protesting Model T Ford chassis. This down-at-heel appliance, which had to be propped up while pumping to stop the springs from breaking under the load, was still in use at the end of the 1930s . . .

At first, it was the Merryweather *Fire King* which dominated the self-propelled fire engine market: as developed, this design was available in six basic forms, ranging from a 300-gallon model costing £850 to a majestic 1000-gallon model at £2275. Coke or oil-fired boilers could be specified and the machine carried 100 gallons of water, sufficient for about half-an-hour's running plus enough fuel for four hours' operation. *Fire Kings* were exported all over the world – a special export option was the fitment of iron-spoked wheels instead of the standard heavy wooden wheels to combat the ravages of termites – and the model was still available to order in 1918. The chief of the Singapore Harbour Board Brigade was such a *Fire King* enthusiast – his company operated two, which could be started from cold within 40 seconds – that in 1922, after the model had finally gone out of production, he ordered the nearest available equivalent, a Greenwich Gem steam pump mounted on a Merryweather petrol chassis. It was noticeable that, though the pump was at the very rear of the vehicle, the engine had an exaggeratedly long wheelbase, no doubt to keep the flames of the boiler as far away from the petrol tank as possible.

Shand Mason, who had tried – and failed – to build a self-propelled engine as early as 1877, did not attempt to compete with their ancient rivals. They stuck rigidly to the steam engine and slid into an inevitable decline: in 1922, Merryweather took them over, and sold off their stock at knock-down prices. Perhaps their most fitting epitaph was the parting shot from a Merryweather pamphlet of the 1890s, highly critical of Shand Mason's habit of building special fire engines for competition work: 'A long lived engine', sniffed the author of the pamphlet, almost certainly 'J. C. M.' himself, 'is always to be preferred to a "racer".'

Other European manufacturers were soon following the Merryweather lead: in 1904, the Paris fire department put a Weyher et Richemonde steamer into service; this was built on similar lines to the *Fire King*, and had a 30 h.p. compound steam engine which could be 'simpled' to give more power for starting or hill-climbing. It had a pumping capability of 575 gallons per minute, and could travel at 15 m.p.h. carrying a crew of twelve men.

In Hanover, Germany, Director Reichel, who led what was claimed to be one of the country's most up-to-date fire departments, was an ardent advocate of the self-propelled

(Opposite) A Fire King on test in about 1905: this engine was destined for the Army at Aldershot

(Above left) Testing a Merryweather Fire King on Blackheath, 1905

Firefighting reduced to basics – pedal-powered Merryweathers of 1886 (right) and 1895 (left) carried hose which could be coupled to the mains to keep a fire under control until larger engines arrived

A Fire King in action, from the cover of a 1903 Merryweather catalogue

fire engine, but felt that the short delay necessary while a steam boiler heated up might be sufficient to make the difference between a fire's being controllable or out of hand, and so he drew up a set of requirements for a steam engine of novel design for his brigade.

1: The engine is to be run out at once after the alarm is given.
2: It is to travel without any noise.
3: During the run it is not to give off smoke or odour of any kind.
4: When it reaches the fire it is to have full steam pressure, ready to deliver the water stream at once.

A somewhat complex procedure was laid down to comply with all these criteria. While the engine was standing in the station, a small gas-burner kept the boiler water warm; however, so that the engine could be driven while pressure was being built up in the boiler, three cylinders of carbonic acid gas were carried. The gas pressure was enough to keep the motor turning over so that the engine could be run on the road while the boiler was being fired up.

The engine was silenced by a condenser connected to the exhaust pipe; though it can have made little noise even without this fitment.

Gas pressure was also used to spray alcohol into the furnace underneath the boiler, so that working steam pressure could be generated while the engine was on its way to the fire: this took about ten minutes, after which the unfortunate engineer had not only to change over from gas propulsion to steam power, but also to switch fuels, throwing 'smokeless charcoal briquettes' into the furnace, where another alcohol spray ignited them. Once the charcoal was burning, *another* fuel supply had to be brought into action, and the furnace was duly charged with 'peat-coke'.

'Thus', concluded the Herr Direktor proudly, 'when the engine arrives at the fire, it is prepared to deliver steam to the pump, the firing being continued by the use of coal or coke'.

No doubt Reichel's invention satisfied his sense of Teutonic efficiency, but one imagines that the hapless engineer in charge of the cumbersome engine wished that his chief had adopted the far simpler British idea of keeping steam up while the engine was in the station with a gas ring under the boiler. Incidentally, British steam engine crews always laid their boiler fires upside down: coal or coke at the bottom, then kindling wood, then paper on top. The idea was simple: as soon as the alarm sounded, the engineer took a shovel-full of hot coals from the station fire and threw it into the furnace, so that the boiler had an instant fire and an instant full head of steam.

While the European firefighting industry had been expanding into new fields, the Americans had been closing ranks. There was, it seems, a surfeit of fire appliance manufacturers in the New World, and the largest and best-known had been forming alliances to protect themselves against competition.

It had started in 1891, when the Button Fire Engine Works, Silsby, Ahrens and Clapp & Jones united to create a 'trust' under the corporate title of the American Fire Engine Company. LaFrance, invited to join the combine, politely refused.

But in 1900, five years after the death of Truckson LaFrance, further overtures were made to the company by a group of New York financiers who were attempting to create a fire engine monopoly, and this time they accepted the offer, becoming a component of the International Fire Engine Company, an ambitious organization which linked three makers of steam fire engines (American Fire Engine Company, LaFrance and Thos. Manning Jr. & Company), three support equipment manufacturers (Rumsey, Gleason & Bailey and Chas. T. Holloway) and three fire extinguisher manufacturers (Chicago Fire Extinguisher Company, F. E. Babcock Company and the Macomber Chemical Fire Extinguisher Company). Options on two other firms, Amoskeag and Waterous, were not taken up.

There were plans to establish a corporate factory in Chicago Heights, Illinois, but though a plot of land was purchased, the group collapsed from lack of funds in 1903. A drastic reorganization and a complete change of management saved the most important constituents of the International company, which reappeared as the American LaFrance Fire Engine Company, a name which was to become virtually synonymous with the manufacture of fire engines in the USA.

The year of the reorganization, Asa LaFrance built a prototype of an aerial ladder raised by a powerful spring; this replaced the old screw-elevated ladders until hydraulics became reliable (the company also experimented with compressed air as a medium for elevating aerial ladders around this time, but though some units were sold, the problems of storing compressed air at fire stations caused the idea to be abandoned).

Soon after the formation of American LaFrance, Chris Ahrens left the company to set up on his own once again: he was soon joined by Charles H. Fox, the Master Mechanic from the Elmira factory which had become the new group's headquarters (the Cincinnati and Seneca Falls operations were transferred here during the period 1905–9), and the two men founded the Ahrens-Fox Fire Engine Company in Cincinnati in 1908.

American LaFrance lost no time in developing a self-propelled steamer: one was in service in New London, Connecticut, in early 1904. Fitted with large chemical foam tanks, the appliance had a separate steam engine to drive each rear wheel.

Testing the hill-climbing powers of the prototype Merryweather Fire King, 1899

Several of these machines were built, and one was exhibited at the Fire Chiefs' Convention in 1905–6, proving itself capable of speeds up to 40 m.p.h.

Another new development at this time was a searchlight wagon, the prototype of which had actually been placed in service in New York City in January 1900 to the order of Fire Chief Croker, who prophesied that his brainchild would 'revolutionise the present method of fighting interior fires by enabling firemen to overcome the great barrier of smoke'.

Firefighting technology was certainly making great strides, but among all these complicated new devices there were still some inventors whose god was simplicity. Like W. A. Dolfini of New York, who in 1899 invented a 'Bicycle Chemical Engine' for four riders.

But Merryweathers had been producing pedal-powered 'first aid' vehicles since 1886, when they converted a two-man tricycle to the specification of Mr Glenister, Chief Constable and Chief of the Volunteer Fire Brigade at Hastings, Sussex. Their 1895 'Quadricycle' hose-carrier consisted of two tandem bicycles linked by a hose-carriage.

The ultimate machine of this type was surely the Truffault Quadricycle Fire Engine, built under the patent of an inventor named Lockyer. The Truffault actually had a pump linked to the pedals, which it was claimed could send a jet of water shooting 80 to 100 feet into the air.

INTERNATIONAL ADVANCES IN DESIGN

One of the earliest Merryweather motor fire pumps dating from 1905: in the background is a chemical escape of about 1910 vintage

The citizens of Finchley, to the West of London, were somewhat perturbed by their borough's vulnerability to fire. Even by the standards of 1903, their local brigade's engine was something of a museum piece, and its value in case of an emergency was questionable. The principal problem, complained Chief Officer Sly, of the Finchley Fire Brigade, was actually getting the engine to the fire. The horses, it seems, had to be borrowed from a local stables, and it was often half an hour before they could be brought to the station and harnessed up, by which time the fire might well have got out of control. The other problem was that the local water company's mains operated at a very low pressure, so that a powerful pump was needed.

So, in January 1904, the local council sat down to deliberate Mr Sly's request for a petrol motor fire engine. There were few precedents to give them guidance: Cambier's pioneering venture had been quite forgotten, and the only other internal-combustion-engined appliances in use were only first aid machines, not pumpers. The Liverpool Brigade had tested a Daimler fire tender in the summer of 1901, but its reliability record was poor, and its temperamental hot-tube ignition had caused much backfiring, earning the machine some ribald nicknames. Then Eccles in Lancashire had acquired a little 7 h.p. Bijou first-aid tender from a local manufacturer, but this was little more than an underpowered toy. The most successful of those early petrol appliances had been the Merryweather delivered to Tottenham Fire Brigade in 1903, a combination fire-escape and chemical first-aid engine. Powered by a 20 h.p. Aster engine, the Tottenham Merryweather could travel at 15 m.p.h., and could turn out in under 20 seconds. Indeed, the Tottenham Brigade had shown their confidence in the future of the self-propelled fire engine by opening a new fire station at Harringay which had no provision for horses.

SCIENTIFIC AMERICAN SUPPLEMENT. No 1883

Charles Allen Munn, President, 361 Broadway, New York.
Frederick Converse Beach, Sec'y and Treas., 361 Broadway, New York.

Entered at the Post Office of New York, N. Y., as Second Class Matter.
Copyright, 1912, by Munn & Co., Inc.
Published weekly by Munn & Co., Inc., at 361 Broadway, New York.

NEW YORK, FEBRUARY 3, 1912.

Scientific American Supplement $5 a year.
Scientific American and Supplement, $7 a year.

Scientific American, established 1845.
Scientific American Supplement, Vol. LXXIII, No. 1883.

A Martin-Tractor attached to the "water tower."

The Water Tower

A New Type of Horseless Fire-fighting Truck

We have had occasion repeatedly to draw attention to various adaptations of the automobile to fire service. There are a number of reasons why just for this function motor propulsion should be adopted in preference to the old-fashioned horse traction. As yet there has been no standardization in the form of motor vehicles for this branch of public service, and it is therefore interesting to take note of such original designs as are brought out from time to time.

A novel vehicle of this kind has just made its appearance on the streets of Springfield, Mass., which creates as much excitement as did the first horseless carriage some years ago. It consists of a piece of fire-fighting apparatus, known as a water tower, with a motor attachment in front, to take the place of horses.

The device is built on a steel frame carrying a motor and transmission machinery, which is placed in front of a fire engine or other apparatus originally designed to be horse drawn. It rests on the front axle of the vehicle to which it is attached. In front the structure is supported by a single wheel or two wheels placed close together. This arrangement gives it a more or less freaky appearance, but is said to be necessary in order to make short turns and handle the vehicle in narrow streets.

The frame is supported on easy springs both in front and in the rear, which prevents the engine and drawing mechanism from being shaken to pieces on rough roads. The engine is geared by means of chains and sprockets to the front wheels of the vehicle, the whole making a complete motor-drawn piece of fire-fighting apparatus.

This device is adapted to all kinds of heavy trucking as well as it is to fire apparatus.

Mr. Martin, the inventor, speaking of the machine, says: "It proved in practice all it promised in theory, and that is all we could ask. We are more than pleased with the tests. We carried 18,000 pounds up an 8 per cent grade on muddy road surface with only 21 per cent of the total load on the driving wheels, which were fitted with special traction tread steel tires. There was not the slightest slippage, nor could we make the wheels skid on the wet asphalt, proving conclusively that it is possible to obtain traction with a steel tire.

"We then drove over curbs, through plowed fields, through sand banks, maneuvered through the woods,

Making a complete turn in a thirty-foot street. Tower and Tractor measure 57 feet over all.

Scientific American

January 11, 1908.

fire, preventing it from feeding upon the oxygen. There are so many possible combinations which could be utilized for the same purpose, either in the form of liquid or gas, that some inventor who is an expert in chemistry should give a substitute for water, or at least a liquid which would be a powerful auxiliary to it. The elements we have named as well as others, can be obtained in such quantities that the cost of such a composition should not prevent it from being prepared in sufficient quantities to be used on a large scale. Experts agree that a gas or a liquid could be manufactured and stored safely in reservoirs properly protected, and that no reason exists why the compound should not be distributed through piping, possibly ejected by means of powerful pumps, as in the case of high service water systems. The modern hose would be strong enough and sufficiently impervious to carry the gas or liquid, so that it could be utilized at a fire in the same manner as water is now employed.

Apparently the principal problem to solve is that of the chemist in obtaining a solution which, while suitable as a fire extinguisher, will not be so affected by low temperature as water, and will accomplish much greater results in proportion to the quantity used. When we remember that a pound of wood contains enough heat units to evaporate a gallon of water, it is not strange that in conflagrations a score or more engines may throw 25,000 or 30,000 gallons of water every minute upon the fire yet have no appreciable effect, on account of the intense heat caused by the large quantity of inflammable material which may be afforded merely by the floors, window casings, and interior finish of the so-called fireproof building.

The enormous saving which would be effected by a practical substitute for water, or one which could be employed on a large scale in connection with it, can be appreciated when the fire loss in recent years is considered. The series of great fires beginning with that of Chicago and ending with the San Francisco disaster caused a loss aggregating no less than six hundred billion dollars. It is a fact that within the last twenty-five years the total fire losses have amounted to three and one-half billion dollars, or nearly one million more than the greatest public debt which the United States has ever incurred at one time. Surely there is an opportunity for the chemist to give practical help by improving upon the element upon which we now depend.

According to L'Electricien, a Vienna firm has recently placed on the market brushes made of glass, which are to replace emery cloth for cleaning and polishing the commutators of dynamos and motors. These brushes are said to clean the commutators without scoring the metal, and their use avoids the inconvenience and dangers of emery cloth.

Zero Weather the Water Freezes as It Falls, Covering the Ruins with Thick Layers of Ice.

The Hoses Themselves Are Often Incased in Ice.

Often Left Standing, and These at Times Slightest Warning.

WINTER FIRES.

When inclined at an angle of 76 Degrees and Extended to Its Full Length of Eighty Feet, this Ladder is Capable of Sustaining a Weight of 500 Pounds at Its Upper Extremity.

Firemen's Ladder Extended Directly from Its Supporting Motor Truck. The Ladder Which is Shown in Near View in the Illustration Above is Here Seen in its Extended Working Position.

LATEST EQUIPMENT OF THE PARIS FIRE DEPARTMENT.

Early press reports of motor fire engines: the three-wheeled Martin Tractor was intended as a direct replacement for horse traction

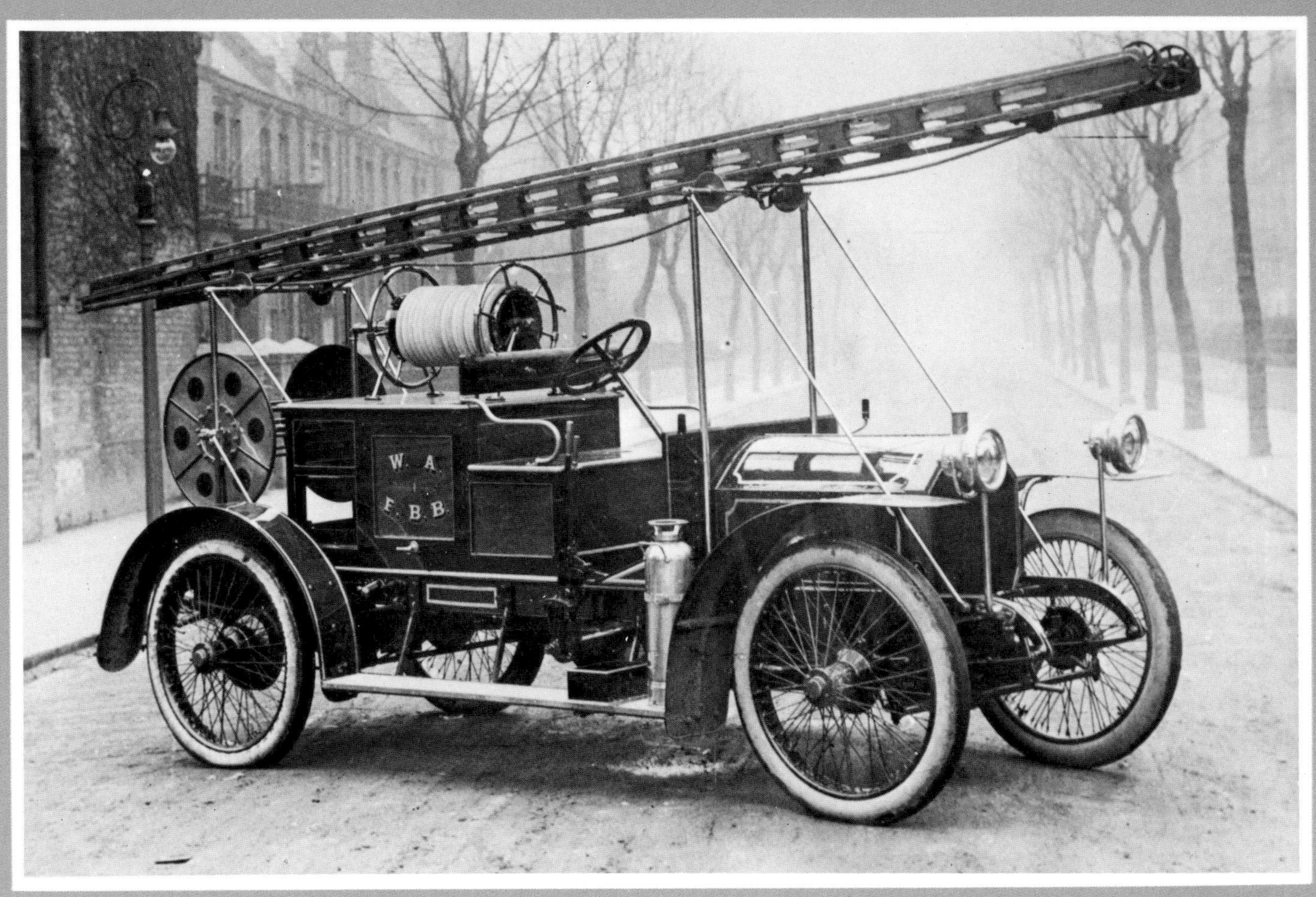
W. A.
F. B. B.

METROPOLITAN
FIRE BRIGADE

But Mr Sly was proposing something even more ambitious: he wanted a petrol-driven fire engine incorporating a pump, an entirely new type of machine for England. Even the prestigious Metropolitan Fire Brigade only had first-aid motors: a Locomobile steam car bought in 1901 for £250, a little Wolseley hose tender and a Merryweather petrol-powered chemical engine delivered in December 1903.

The Finchley Borough Council duly met to consider Mr Sly's request. They were told that the project had been put out to tender, and that Merryweather had offered to supply such an appliance for £980, less an allowance of £60 for their old escape. The nearest quote to this had been £1130 from the Hercules Company of Leamington.

Some mechanophobe members of the council were duly sceptical of the worth of the entire project. Their spokesman, Councillor Warden rose to his feet, and fired a loaded question at the unsuspecting Clerk of the Council: 'What do you reckon is the life of a motor car?'

'About ten years', replied the Clerk, thinking quickly.

'But you have no guarantee that it will last ten years!' riposted Warden. 'They have not been in existence that length of time'. And sat down, doubtless to the accompaniment of the delighted sniggers of his supporters . . .

However, the Merryweather tender was accepted, and subsequent events were to prove Councillor Warden sadly wrong.

In fact, before the Finchley Merryweather arrived, another motor pump was built by this manufacturer and exported to the French estates of Baron Henri de Rothschild in June 1904. Like all the early Merryweather motors, it had a chassis built by Aster of Wembley, who specialized in supplying running gear for other manufacturers, who just had to add bodywork and their nameplate to claim it as all their own work. With a Hatfield pump, the Rothschild Merryweather could pump 300 gallons a minute to a height of 120 feet.

A similar pump was fitted to the Finchley engine, which was given its acceptance trials in November 1904 at Christ's College, Finchley, in front of an enthusiastic crowd. Within seconds the escape ladder was removed from the engine and run up to a second floor window, while the chemical hose which was fitted as an auxiliary to the motor-driven pump was brought into play as a suction hose was taken to the college swimming pool. Then the motor pump was set in motion, and a fine jet of water around 150 feet high was thrown against the college tower. Powered by a 24/30 h.p. four-cylinder Aster engine, the Merryweather could attain a speed of more than 20 mph on the level. Its fire crew, who had to travel standing on footboards at either side of the machine, were doubtless grateful that it could go no faster: but in 1912 the Merryweather was modified to increase its hill-climbing ability, and given a

(Opposite, top) A Napier chassis of circa 1911 equipped as a fire engine for Western Australia

(Opposite, below) Merryweather's factory yard at Greenwich, circa 1905, with motor fire engines in various states of completion: in the background, under a tarpaulin, is the Sutherland steamer of 1863

(Above) The first petrol fire pump, built by Daimler in 1888

Fire drill with a Merryweather escape at London Fire Brigade headquarters, Southwark, around 1910

new 50 h.p. Aster engine. In 1913, the Finchley brigade acquired a more up-to-date fire engine, and the old Merryweather was relegated to stand-by duties: the brigade, with notable lack of sentiment, sold it in 1928 for £25 to a Hounslow gravel merchant who used the engine and pump for washing gravel from his pits for a couple of years, and then disposed of the remains to a scrap dealer in Fulham. Then the London Science Museum were informed of the machine's peril, bought it and restored it: this Merryweather now stands alongside another historic fire engine, the old *Sutherland* steamer of 1863.

Not every brigade had such reliable service from their motor fire engines as did Finchley: in 1904, Brighton bought a petrol appliance which was nothing but trouble. The local newspaper catalogued some of its adventures:

> Shortly after the engine was acquired by the Corporation, it caught fire when proceeding along the Old Stein for the purpose of extinguishing a house-fire . . . It was eventually returned to London for a complete overhaul . . . Superintendent Lacroix visited the Metropolis to bring it back by road. At Hazeldean, about fourteen miles from Brighton, the steering gear went wrong, and the engine was precipitated into a ditch after toppling over a van.

With the petrol fire engine giving such ambiguous results, it's not surprising that other methods of propulsion had their advocates, like Superintendent Alex Weir of the Liverpool Brigade, who commented:

> The self-propelled fire engine is undoubtedly the engine of the future and, inspiring as the sight of galloping horses dashing along the streets with the fire apparatus undoubtedly is, and much as it will be missed by the public in general, yet it will certainly pass away in response to the call for more speed at less cost, or in other words, higher efficiency with greater economy.
>
> There are three methods of propulsion to choose from: steam, electricity and petrol. Taking the latter first, my personal opinion is that it is the least reliable.
>
> A perfect first turnout fire engine is an electrically-propelled vehicle of sufficient capacity to carry a good supply of hose and tools, and a good stock of extending and single ladders . . . and fitted with a 60 gallon tank, the water from which is forced out by carbonic gas, carried in a cylinder.

Perhaps 'perfect' was the wrong word to apply to the battery-electric vehicle, which enjoyed a brief vogue at the turn of the century. Such a vehicle could certainly be called out immediately, with no starting problems – and with no fire risk from a tank full of petrol – but was dependent on a vast weight of short-lived batteries which needed recharging at frequent intervals.

Equally, the battery electric *had* to be a chemical engine, for it had no means of driving a pump: most engines of this type had

(Left) In 1905 Merryweather demonstrated a light pump that could be driven by a jacked-up touring car, in this case a Charron, Girardot & Voigt

(Below) A 1906 German self-propelled steamer

A horse-drawn steamer in action during the San Francisco earthquake of 1906

Aftermath of the disastrous fire caused by the 1906 San Francisco earthquake

their motors in the hubs, like the Cedes (built by the Austrian Daimler Company) which was perhaps the most widely used appliance relying on batteries.

To overcome the inherent shortcomings of the battery electric, various hybrid engines were devised, though these posed more problems than they solved: among the combinations which enjoyed a limited vogue were petrol-electrics, in which the petrol engine worked the pump and drove a dynamo supplying current to the traction motors, steam-electrics, with electric drive and steam pumps, and petrol-steam. Then there were the dual engines, which duplicated their motive power, using one engine to drive and one to pump.

Each one of these types saw service with a brigade somewhere in the world: the most famous builder of petrol electrics was Löhner-Porsche of Vienna. The Viennese brigade, however, apparently preferred steam-electrics. The Nuremberg fire engine builder Justus Christian Braun was one of the most persistent manufacturers of the latter pattern of engine.

In 1909 the Hamburg Brigade decided on the steam-electric combination, too, though their means of attaining this end was far simpler than buying a new engine: they merely hooked one of their old iron-shod horse-drawn steamers behind one of the latest Mercedes Electrics, which was capable of hauling it at claimed speeds of between 20 and 35 m.p.h. It was further claimed that the mounting of the electric motors directly on the wheels 'avoided any risk of the driver losing control of the car in case the current supply should be discontinued, as no critical momentum could arise'.

Another German city, Breslau, favoured the petrol-steam combination, which was also

used, albeit on canal-going fire appliances, by Venice. But perhaps the most curious combination of all was that adopted by the Hanover Brigade in 1911, and once again that purveyor of freaks Herr Braun was the supplier. This curious device – in fact, Hanover had two – had a battery-electric chassis with a range of only 30 miles, but used a rear-mounted petrol engine of 54 h.p. to drive a Pittler fire pump.

In America, the art of building curios was carried to a high degree of perfection, for several makers supplied two-wheel motorized front-wheel drive 'power packs' to convert old horse-drawn equipment to self-propulsion: LaFrance even made a six-cylinder unit of this type, which developed 105 b.h.p., and which must have been quite hair-raising when coupled to an old steam pumper.

But LaFrance's entry into the field of motorized fire engines had been quite conventional – forced on them, indeed, by changing fashions as, towards the end of the first decade of the twentieth century, fire departments all over America, fired with the 'hustle' that was revolutionizing that country's technology, turned from the steam fire engine, self-propelled or horse-drawn, and demanded that all their new units should be petrol-powered.

LaFrance first experimented – unsuccessfully – with a petrol engine of their own design, then decided to take the safe course of buying in a proven power unit. They purchased the very best that America could provide – a specially built Simplex chassis, constructed by a maker normally associated with high-class luxury cars. Based closely on the contemporary Mercedes, the Simplex had cylinders and pistons of the finest 'gun iron' and a chassis frame pressed from Krupp's chrome nickel steel. Designer Edward Franquist created a special four-cylinder engine for the LaFrance company, who endowed this

high-powered chassis with 'combination' (chemical and hose) bodywork.

The new motor fire engine was demonstrated at a Fire Chiefs' Convention in 1910 in Syracuse, New York, and achieved a speed of 50 m.p.h. on a race track at the local fairground. The first production petrol-powered LaFrance combination fire engine was sold to the Lenox, Massachusetts, Fire Department the same year: it is still preserved in the Lenox firehouse in running order. The second engine of this type was sold to Dallas, Texas.

It seems as though the Simplex engine was not entirely suited to its new role, for LaFrance now set about designing its own six-cylinder engine specifically for fire-fighting vehicles: but the cost of all this redevelopment exhausted the company's financial resources, and a new, better capitalized company was incorporated in January 1913. Experiments were also made in providing other types of vehicles to fire departments, and the firm built a number of Fire Chief's cars, one of which bore the evocative name *Gray Ghost*.

Despite the early hopes of a fire engine monopoly, LaFrance were by now far from being alone in the fire engine market. Two of America's more eccentric inventors, Walter Christie and Harry Austin Knox, provided conversion tractor units for fire departments. Christies, who had a life-long obsession with front-wheel-drive, created a fearsome two-wheeled device of massive appearance, while Knox specialized in three-wheeler tractors, with the single front wheel ahead of the brass radiator. This type of tractor (a similar device by another maker was known as the Martin-Tractor) was a forerunner of the modern articulated vehicle, a coupling on the back of the tractor replacing the front axle perch of the formerly horse-drawn trailer. 'This arrangement', noted the *Scientific American* drily, 'gives a more or less freaky appearance, but is said to be necessary in order to make short turns and handle the vehicle in narrow streets'. Springfield, Massachusetts, were so enamoured of their Knox outfit that they kept it until the Second World War.

In addition to the tractor unit, Knox also built complete fire engines, one of which was sold to a South American municipality on condition that the machine would give a satisfactory demonstration of its capabilities before the order became binding. On demonstration day, the engine was duly connected to a water hydrant, and began pumping in fine style; suddenly the fierce jet of water died to a trickle . . . the pump had drained the local water mains dry, and the sale was off!

Knox may not have lasted long in the fire engine business, but several marques who entered the field around this period were laying the foundations of continuing business. Take the Mack, built by a determined family of five brothers, who in 1900 had built what was claimed to be America's first motor bus: in 1909 they produced a motorized hook-and-ladder truck for the brigade in Morristown, New Jersey, who kept it in service until 1926. In 1911 Mack built their first pumping engine, which also incorporated the first production example of a new type of rotary pump called the Gould, distinguished by an exceptionally large air chamber. This engine was sold to Cynwyd, Pennsylvania; similar units were sold the following year to Baltimore, Maryland, Morristown, New Jersey and New York. It was the start of a period of growth which was to see Mack become one of the biggest manufacturers of fire appliances in America.

(Right) Primitive breathing apparatus used in America around 1910

One Mack model became legendary, the ruggedly ugly Model AC, introduced in 1915, which was nicknamed the 'Bulldog': many different types of fire appliance were built on this chassis, which was also adopted as the standard 3½/5-ton US Army truck during the First World War. One peculiarly American appliance based on the AC Mack was an articulated aerial ladder unit of such inordinate length that the rear axle had to be made steerable, and put under the charge of a separate steersman, who sat in solitary splendour at the back of the apparatus to guide his end of the machine round sharp bends.

Another classic model, which first appeared in 1911, was the Ahrens-Fox, for many years distinguished by its massive pump, surmounted by a circular air vessel, mounted above the front axle with the square-cut radiator and bonnet set well back, imparting a purposeful air to the appliance. Though the quality of their engines was above reproach, in later years Ahrens-Fox had a somewhat chequered career, eventually being taken over by Mack in 1956, and vanishing from the scene shortly afterwards.

The Seagrave Company was founded in 1907 in Columbus, Ohio, and its early

(Left) An Edwardian Hotchkiss fire engine still preserved in running order in Australia

(Below) This 1912 John Morris-Belsize had an 85 ft. turntable ladder actuated by compressed carbon dioxide

An early Merryweather chemical motor fire engine delivered to the Willesden (London) Brigade

petrol-engined models, one of which still survives, were chemical engines with forward control and underfloor engines, a layout which was advanced in concept, if not in execution, but by 1914 the company was building long-bonneted engines of more conventional appearance. Other makers of forward-control fire engines at this period were Packard and Pope-Hartford, but these were really no more than aberrations from their normal output of luxury motor cars, and did not represent a continuing line, as did the Seagrave.

One of the longest-lived American fire appliance manufacturers is Peter Pirsch & Sons, whose origins can be found in the nineteenth-century Nicholas Pirsch Wagon and Carriage Plant in Kenosha, Wisconsin, where in 1899 Peter Pirsch devised and patented a compound-trussed extension ladder, starting his own company the following year to exploit this and similar firefighting devices. The firm's first venture into the self-propelled fire engine field dates back to 1910, when a 'combination' chemical engine was produced. Only chemical engines were produced for the next few years, but in 1916 the Pirsch company produced a 'triple combination pumping engine' on a petrol chassis for the city of Creston, Iowa, as well as a similar machine for Chicago.

These were the marques which earned a lasting fame: more transitory success was enjoyed by such manufacturers as the Webb Motor Fire Apparatus Company, which was run by a former motor-racing driver named A. C. Webb, who in 1907 converted a Thomas Flyer touring car (similar to that which won the 1908 round-the-world New York to Paris race) into a fire engine, with a rotary pump driven from the transmission. Initially based at Vincennes, Indiana, and later at St Louis, Missouri, Mr Webb's venture had a life span of only four years.

In 1908 Henry Ford had introduced his 'Universal Car', the immortal Model T, which was to remain in production for nineteen years, during which well over 15 million cars and trucks were built on this chassis; the two-seated Model T Runabout car proved popular as a fire chiefs' car, starting in 1911 when a special series of ten Torpedo Runabouts finished in fire engine red was built. The standard equipment on this 'Chief's Model' included a rear locker for a fire extinguisher, a coil of rope and other firefighting implements; a brass fire bell (bells, incidentally, only became normal wear for fire engines in the early twentieth century, though a 'celebrated Australian pugilist' had tried to sell the idea to Merryweathers as early as 1890); a front bumper; detachable rims and a spare tyre 'inflated for emergency use'. Fire Commissioner Waldo of New York City was one of the ten fire chiefs who took delivery of one of these cars.

It was not for several years after this that the Model T chassis became generally used as a

basis for fire engines, however, as in its standard form it was too short in the wheelbase. One of the very first fire engine conversions of the Model T to appear was actually built in Britain in 1916: it was a one-off truck, carrying a generous supply of chemical extinguishers optimistically intended to control fires on the airship base at which the engine was located. Just how eight or ten chemical extinguishers were expected to deal with a blaze in an airship several hundred feet long and filled with highly inflammable hydrogen was not specified.

In America though, firefighting was taken into an entirely new dimension with the very first flying fire engine, which went into service in the summer of 1918.

San Diego's 'Aerial Truck No 1' was in fact a Curtiss hydroaeroplane equipped with two 3-gallon chemical extinguishers and four carbon tetrachloride extinguishers; carrying a crew of two, it had a 110 h.p. six-cylinder engine and was capable of 70 m.p.h.

Despite this pioneering effort in San Diego, the age of the 'hydro-fireplane' was, in reality, still a very long way in the future: back in 1909, however, the American *Fireman's Herald* had published a prophetic drawing of 'Fire-fighting in the New Age', depicting a Wright biplane equipped with hoses fighting a fire in a skyscraper. It was inspired by the fact that already America's tallest buildings were four times higher than the reach of the biggest escape ladders then available.

(Top) Around 1910, Hitchin Brigade took the unusual step of adapting a powerful touring car to pull their old Merryweather steamer

(Above) A Daimler chassis, circa 1909, converted into a chemical fire engine

(Left) Merryweather produced this type of fire engine right into the 1920s

TECHNOLOGICAL DEVELOPMENTS

(Above) A Maginus turntable ladder at a 1920s display in Birmingham

(Opposite, above) Merryweather built this Warspite airfield crash tender in 1935 on a Morris-Commercial chassis

(Opposite, below) The London Salvage Corps in action at a fire during the 1920s

At the beginning of the 1920s Merryweather, which had for many years dominated the British fire engine market, found itself with two rivals, both of whom had entered the field by chance. The Dennis Brothers, John and Raymond, of Guildford in Surrey, had been bicycle makers during the 1890s (their 'production line' at that period consisted of the branches of a tree behind their shop on which cycle components were hung as the machines were assembled) before turning to motor cars. They had subsequently developed an excellent shaft-driven commercial vehicle chassis – their list of customers ranged from the balloon manufacturers Paddon & Sopwith to the Metropolitan Asylums Board – on which, in 1908, they decided to build a fire engine, using a pump supplied by Gwynnes of Chiswick. This was a centrifugal pump, in which a rotating vane, rather than pistons, developed the pressure, and the unit proved so successful on test that Dennis had no difficulty in selling it to the Bradford Fire Brigade. Encouraged by this sale, Dennis began to concentrate more and more on the manufacture of fire engines, acquiring the rights to a powerful turbine pump designed by an Italian engineer named Tamini. At first, Dennis used Aster engines, like Merryweather, but eventually turned to the White & Poppe engine, to which they finally bought the manufacturing rights. By the time that the Great War broke out, Dennis were firmly established as suppliers of fire engines to the London Fire Brigade.

Leyland's introduction to the fire engine market had been even more casual. Established as steam engineers, the company had first gained fame with fearsome steam-powered lawn mowers, the first of which was supplied to Rugby School in 1895. In 1896 came the first of many steam wagons, then in 1904, an experimental petrol truck nicknamed *The Pig*. The shortcomings of this chassis were speedily rectified, and a reliable petrol-engined truck became available alongside the steamers (though the 1907 production total of 36 steamers against 17 petrol chassis shows both the size of the commercial vehicle market and the prejudice against internal combustion existing at that period).

However, in 1909 the Chief Fire Officer of Dublin arrived at the Leyland works, and told the astounded company that he had decided that they built the finest petrol truck chassis in the country, and that nothing short of the best would be good enough for his proposed new engine. Leyland protested that they knew nothing of fire engine design, but the astute fireman had come armed with his own ideal specification; and so Leyland agreed to build his engine.

It achieved notoriety on test: while the pumping capabilities of its 250 gallons per minute Mather & Platt turbine pump were being tested in the Leyland workshops, workmen began to keel over and fall to the floor unconscious. That was how Leyland discovered the noxious effect of exhaust fumes in a confined space.

The engine was delivered to Dublin in 1910: on trial in Phoenix Park it attained 60 m.p.h., no mean feat on solid rubber tyres. A banquet was held to celebrate its arrival, only to be interrupted by a call to a fire at Kingstown. Driven by Henry Spurrier II, one of the partners in the Leyland company, 'the machine put up fast time on a slippery road, and put the fire out in time for Mr Spurrier to catch his boat back to Liverpool'. The Leyland was the first of many fire engines to be constructed by this maker; it was still in use thirty years after, and indeed was converted to pneumatic tyres at the start of the Second World War to extend its working life. Fire engines quickly became an important part of Leyland's production; late in 1910 a special 85 b.h.p. six-cylinder engine intended for fire engine work was added to the range. Before long, all Leyland fire engines were fitted with the four-stage Rees Roturbo turbine pump. Their fame was widespread: among the early purchasers of Leyland fire engines were the fire brigades of Shanghai and Hobart, Tasmania.

Other British makers whose products achieved a more localized fame were Halley of Glasgow, who supplied a number of engines to the Edinburgh Brigade, and Argyll, whose purpose-built motor factory just outside Glasgow was one of the industrial wonders of the age, who built an impressive fire engine with a turbine pump for Dundee in 1913.

During the 1920s, another Scottish maker was to supply chassis for Merryweather, whose chain-driven fire engines had become quite outpaced by their younger competitors by the mid-1920s. The Albion was a shaft-drive vehicle of up-to-date design, and formed a fine basis for Merryweather's proven pumping apparatus; later, Merryweather also used Ford and Morris-Commercial chassis.

Merryweather were also active in the manufacture of smaller units: they had built a trailer pump in 1906, which could be driven through rollers by the wheels of the towing car, and had developed a petrol-powered Hatfield trailer pump in the 1920s. In 1910 'first-aid' outfits based on motor-cycle and sidecar combinations were pioneered, fitted either with a four-man manual pump or with a standpipe, 500 feet of hose, ladder, tools and

A Dennis engine at a fireman's funeral during the 1920s

Melbourne, Australia, Fire Brigade used this Albion pumper in the mid-1930s

(Left) Leyland engines with Braidwood bodywork used by the London Salvage Corps in the late 1920s

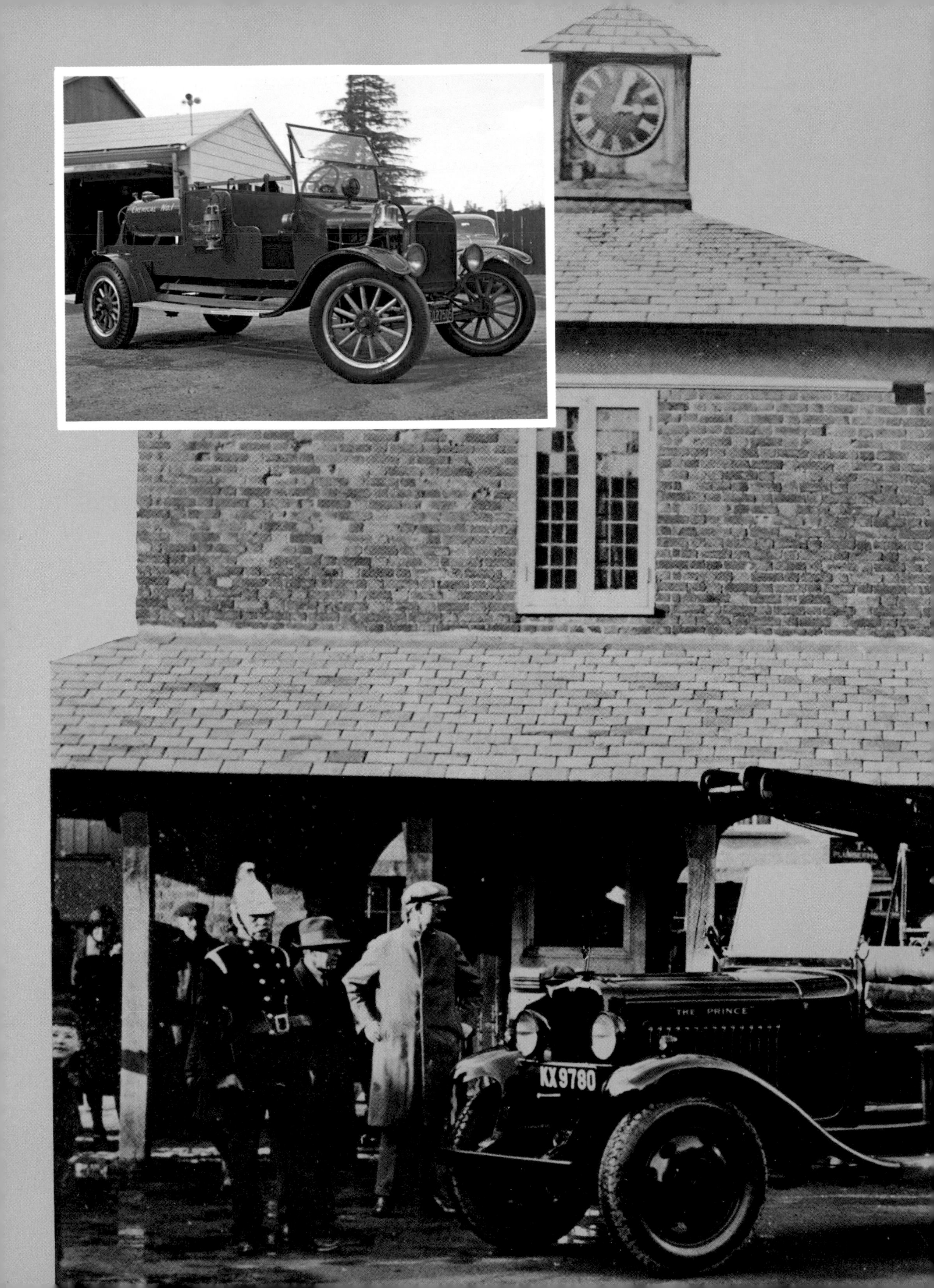

CHEMICAL No.1
"THE PRINCE"
KX 9780

chemical extinguisher; during the 1920s a more refined version of this latter design, with accommodation for two men and an officer as well as 500 feet of hose was produced. Once a standpipe had been connected to a hydrant, the motorcycle could be driven away, with the fireman on the pillion seat paying out the hose over a roller on the rear of the sidecar. Also carried were 'Fire Suds' chemical extinguishers, for use against fires involving petrol, oil, tar, or other inflammable liquids which could not be put out using water. 'Fire Suds' was a froth produced by chemical action, which smothered the flames so that they died from lack of oxygen.

Merryweather's first full-size 'Fire Suds' engine had been exhibited at the Empire Exhibition at Wembley in 1924 and subsequently sold to the Leeds Brigade. Externally, it looked little different from conventional first-aid engines; in the centre of the body were two galvanized steel tanks, each containing 25 gallons of an alkaline solution, and behind these were six lead-lined compartments containing an acid solution and sealed by a lead-sheathed, rubber-gasketed mahogany lid. At the rear of these tanks was another compartment housing an extra supply of 'Fire Suds' charges in metal cylinders.

The special 'Fire Suds' pump had two separate gunmetal chambers – one fed from the alkaline tank, one from the acid – and was driven at engine speed from the gearbox; each chamber supplied its own hose, the two hoses only being united at the nozzle. As the two fluids combined, they formed the 'Fire Suds' foam. By the use of control valves, the alkaline tanks could be recharged once they had emptied: in this way the engine could produce some 1800 gallons of foam before all the acid tanks were exhausted. In 1933, Merryweather announced a new type of chemical engine, in which the exhaust

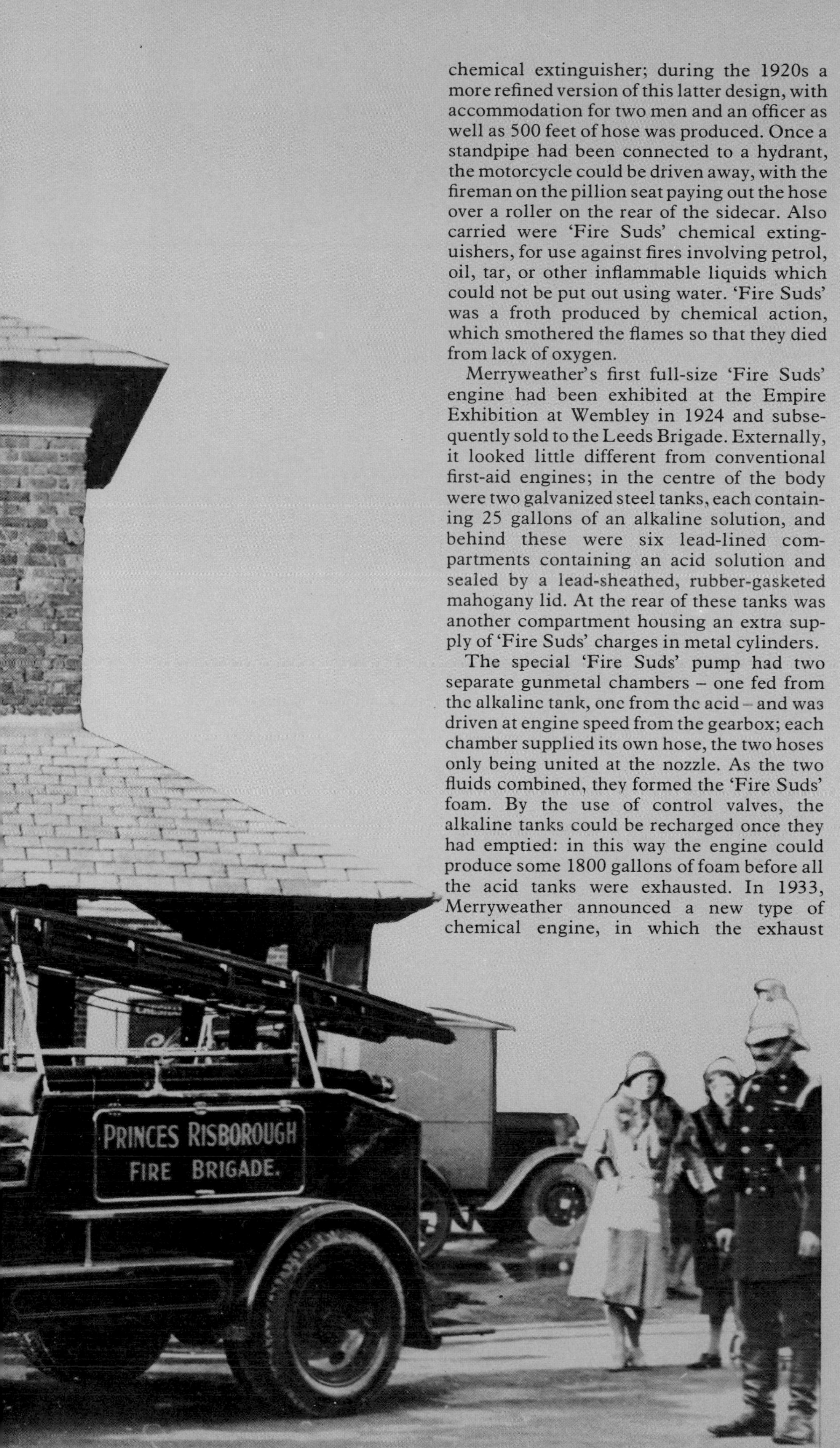

(Top left) Chemical fire engine on a 1926 Model T Ford chassis

(Left) The first Bedford fire engine, supplied to the Princes Risborough (Buckinghamshire) Fire Brigade in the mid-1930s

gases of the vehicle were combined with a 'saponaceous solution' to generate foam.

In America, thought had been given to discovering some means of smothering oil fires as early as 1889, but no suitable foaming agent could be developed at that time: however, in 1909 it was found that extract of liquorice and bicarbonate of soda would combine to form a suitable foam, the only snag being that the cost of processing the liquorice was prohibitively expensive. Then a liquorice importer and processer discovered that a liquorice byproduct was equally as effective as the extract . . . and far cheaper. The company attempted to promote this use of their products, but the process was only imperfectly understood, and many failures resulted. So they determined to enter the firefighting industry on their own account in 1915, and in 1917 established the Foamite Fire Extinguisher Company. A year later they amalgamated with a rival firm, the Erwin Manufacturing Company, to form Foamite Firefoam Company, and were soon selling their products all over the United States. They bought the shells for their extinguishers from the O. J. Childs Company, of Utica, New York, but the recession immediately after the First World War saw Foamite attempting to produce their own shells as an economy measure, but by 1922 they had merged with Childs to form the Foamite-Childs Corporation, based in Utica, and building all types of firefighting equipment from extinguishers to fire engines and ladders. Most of their engines were built on proprietory chassis, though they did make a few chassis to their own design. In 1927 Foamite-Childs and the Fire Gun Manufacturing Company joined American LaFrance, creating the American LaFrance and Foamite Corporation; after the merger, the Utica works were closed down and the Foamite operations moved to Elmira. American LaFrance continued manufacture of the Childs Thoroughbred fire engines for several months after the merger, building 51 of these units before the Childs name was phased out in 1928.

While many American brigades bought new American LaFrance, Ahrens-Fox, Seagrave or Mack appliances during the 1920s, others, less well-heeled, had to make use of touring car chassis adapted to carry firefighting equipment. Most common, of course, was the ubiquitous Model T Ford, and the standard chassis was used as the basis for many a light chemical truck. More ambitious convertors, such as the brigade at Jackson, Tennessee, lengthened the chassis of the Tin Lizzie so that it could accommodate long ladders and extra hose. In Britain, an over-the-counter chassis conversion called the Baico was available, in which the existing rear axle

(Opposite page, top) Leyland-Metz turntable ladder supplied to Macclesfield in the 1930s

(Centre) American La France pumper of the 1930s

(Opposite page, below) Revolution in body design – the Dennis New World

(Below) 1930s rarity: one of only two fire engines to be built on Scammell chassis

was utilized as a jackshaft to carry sprockets for chains which drove solid-tyred rear wheels carred on the extended chassis. At least one of these Baico-Fords was converted into a fire engine, with a Dennis pump mounted ahead of the radiator and driven from the nose of the crankshaft; an awkward installation on this chassis as it interfered with the starting handle, a necessary adjunct to the T's starter-motor in cold weather. The Model T conversions such as this which used a motor-driven pump must have boiled furiously and often as their engines worked overtime; the twin advantages of the Lizzie were go-anywhere robustness and the easy (and cheap) availability of spare parts when the chassis's almost legendary indestructibility was tried beyond its limits of endurance.

If America's lowest-priced car was an admirable basis for fire engine conversion, so, too, was one of the most expensive luxury models, the 13.5 litre Pierce-Arrow 66, the largest-ever American production car, which first saw the light of day in 11.7 litre form in 1910, and survived in production until 1920. It was the ingenious Fire Chief Walter Ringer of Minneapolis who discovered that a second-hand Pierce-Arrow, stripped of its sybaritic coachwork and with its already long wheelbase of 12 feet 3 inches extended still further by lengthening the chassis with channel steel, bolting the existing rear axle to the

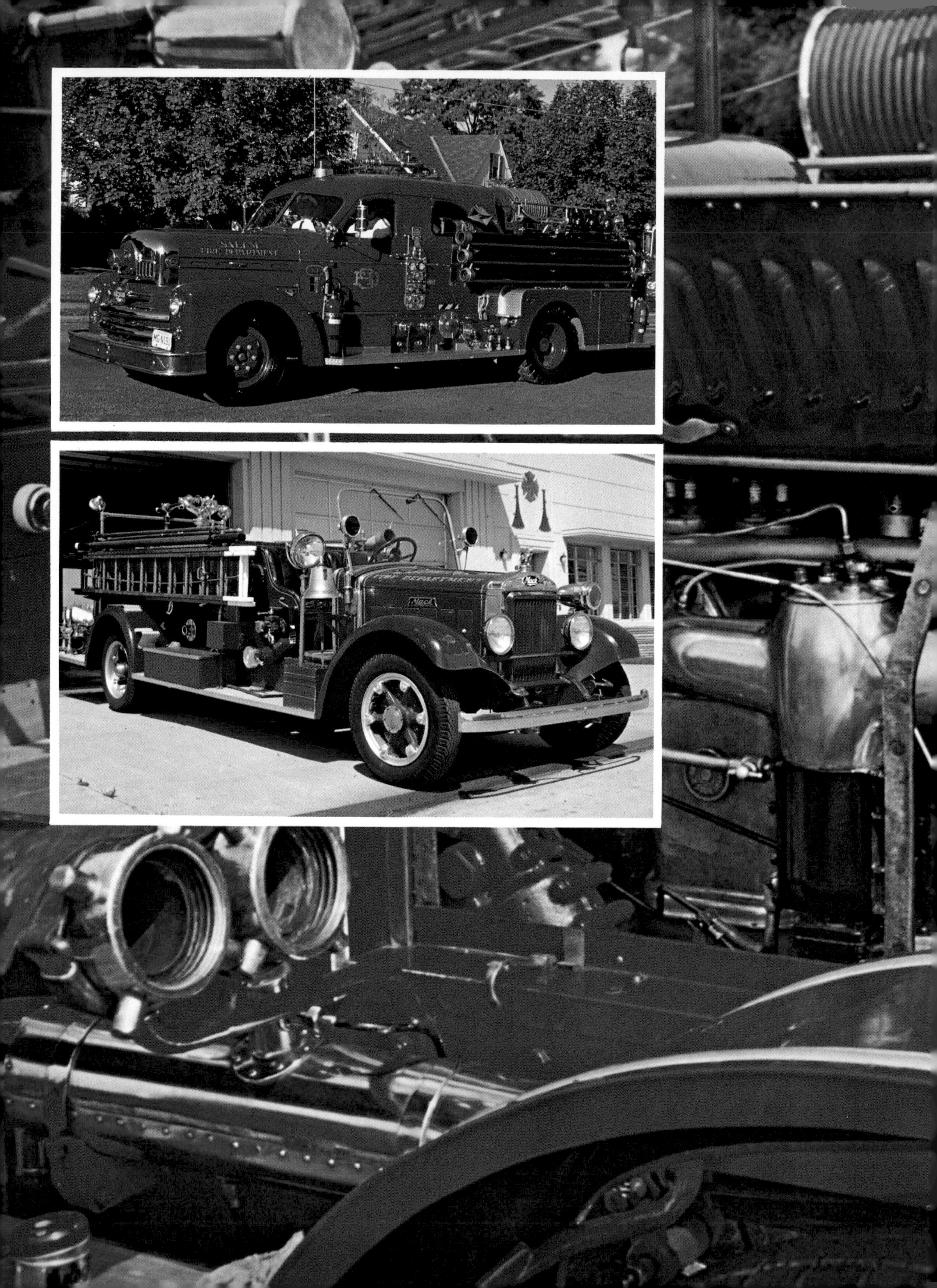
SALEM
FIRE DEPARTMENT
FIRE DEPARTMENT
Mack

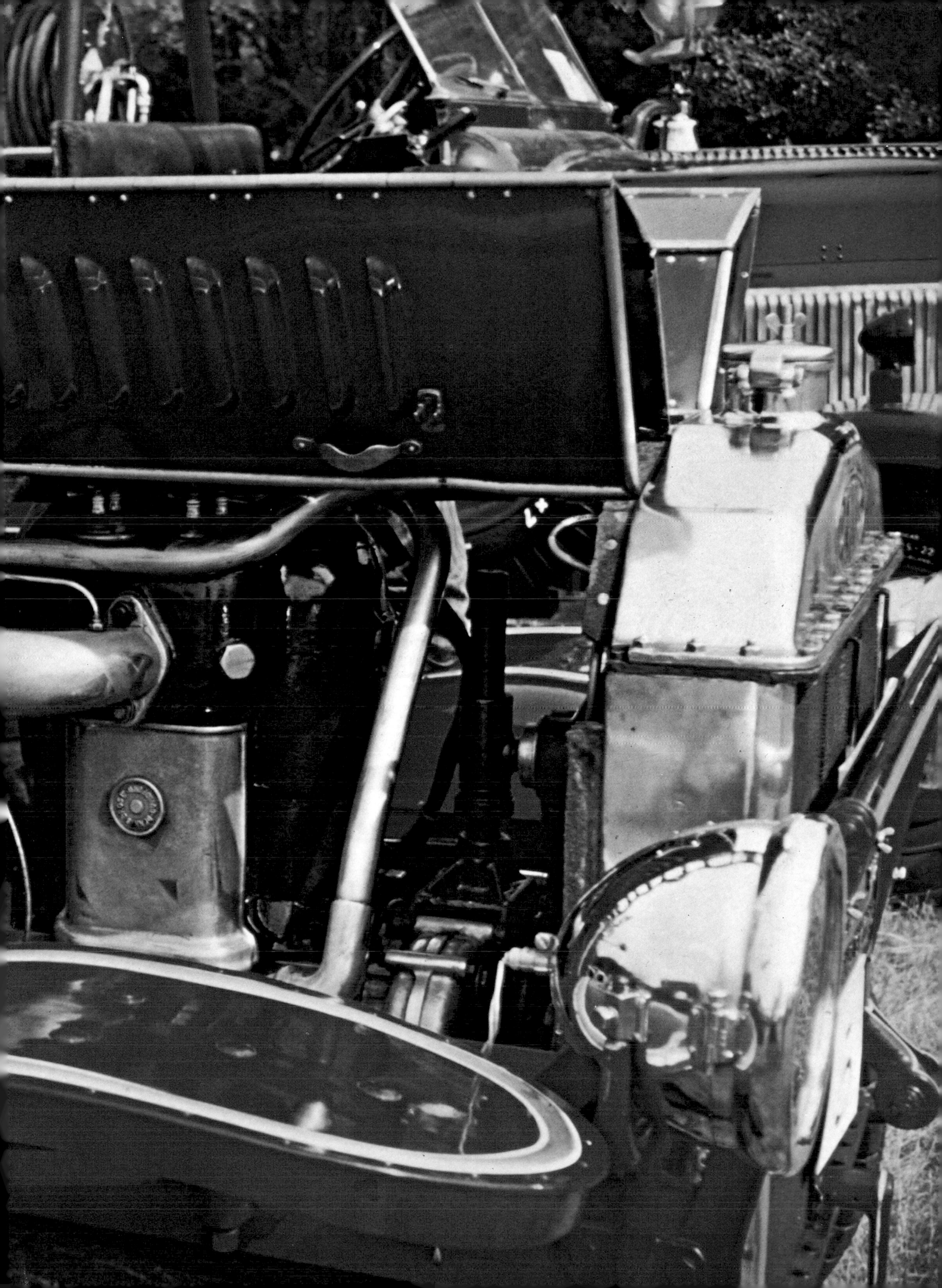

(Previous page) Dennis acquired the White & Poppe company to build power units for their chassis: this is the engine compartment of a 1921 Dennis appliance

(Inset above) Limousine bodywork on a 1954 Seagrave 750 gpm pumper

(Inset below) A 750 gpm Mack pumper built in 1935

frame and using it to carry sprockets to drive the rear axle by chain could be transformed. Ringer's bright idea was widely copied, and quite a number of superannuated Pierce-Arrows ended their days as fire engines, giving as much as twenty years' service in this new guise. And when Pierce-Arrow went into liquidation in 1938, all the engine-building plant which had been used to manufacture the fine new V-12 Pierce power unit announced in 1932 was sold to Seagrave, who adopted this engine for use in its fire engines.

The 1920s and 1930s saw a great flowering of technical development in the design of firefighting equipment: apart from the adoption of foam units, new types of bodywork and improved designs of ladder made their appearance. By now, fire engine speeds had increased so much that the old Braidwood-pattern body, in which the fire crew either sat facing outwards or stood along the body sides, was becoming positively suicidal. Some Continental manufacturers towards the end of the 1920s had already adopted 'toast-rack' bodywork, with the firemen accommodated on transverse benches, but for some reason the British brigades were still using the old-style coachwork designed for the horse-drawn appliances of a century earlier. When in 1928 a fireman named Kidd fell from an engine of the newly constituted Chippenham Centre Fire Brigades' Association and was killed, there was a general outcry in favour of adopting a safer design of bodywork, and the manufacturers looked around for inspiration – and found it in America, where for some time a type of coachwork in which the crew sat facing each other inside the body of the appliance had been in use. The first British manufacturer to deliver a fire engine with this type of bodywork was Dennis – in deference to the origin of the species, it was named the 'New World'. Birmingham and Luton were among the first brigades to take delivery of New Worlds, then, in 1931 Firemaster Methven of Edinburgh commissioned a fully-enclosed 'limousine' pump on a Merryweather/Albion six-cylinder chassis. Despite local opinion that Methven was pampering his firemen, the fully enclosed bodywork meant that the crew arrived at the fire alert and ready for action, rather than weatherbeaten and exhausted, as they had with the old Braidwood bodies. The only snag with the Edinburgh body was that it was rather tall – and therefore topheavy – and eventually overturned in an accident, whereupon the Brigade workshops lopped nine inches from its height to keep its centre of gravity within bounds.

Another 'all-weather' fire engine commissioned during 1931 was built for the Darlington Brigade to the design of Chief W. Spencer: it had a 'rear-entrance saloon' for the crew, carried its suction hose in tunnels down

(Below) Ford chassis were a popular basis for lightweight fire engines during the 1920s and 1930s: this example is built on a Ford BB, current from 1932-35

the centre of the body, and had its pump and controls on a panel at the back of the vehicle. The chassis used for this was the new Dennis Low-load-line with a 5.7 litre four-cylinder engine announced at the November 1929 Commercial Motor Show; with pneumatic tyres all round and servo-assisted four-wheel-brakes, it represented a rapid advance over the typical chassis of the mid-1920s which ran on solid tyres and only had brakes on the rear wheels. (Dennis were also, it seems, the first to offer fire engines with chromium-plated brightwork instead of the traditional brass, thus eliminating a little glamour and a lot of polishing . . .).

More elaborate enclosed appliances were to appear in the early 1930s: in 1932 drawings were published of a new six-wheeled engine for Luton, which looked externally virtually indistinguishable from a luxury motor coach, and carried both a centrifugal pump and a foam-generating mixer, as well as providing accommodation for a crew of eighteen. Even more glamorous was the rescue tender built to the design of Chief Officer Johnson of the West Ham Brigade in 1934:

> likely to become a model for such appliances in large cities and towns, it is entirely enclosed and has a streamlined body mounted on one of Dennis Bros' latest types of chassis with an 80 h.p. four-cylinder prime-mover. There are two rows of seats in front; one for the officer and the driver, and the other for at least five men. Each row has its own doors. In the after portion of the limousine body is accommodation for five rescue-men and their gear, and a fully equipped canteen of novel design.

With the open body also vanished the traditional brass helmets, replaced by cork or leather, which carried less risk of electrocution if they came into contact with a live wire, and were probably even stronger than the metal helmet, which last saw action at the fire which destroyed the Crystal Palace on 30 November, 1936.

Alongside these limousine pumps appeared other new types of appliance, such as the high-speed hose-layer, which carried continuous lengths of hose specially 'flaked' (laid in zig-zag layers to reduce bulk and facilitate paying-out) inside its closed bodywork. In 1936 the London Fire Brigade took delivery of its first high-speed hose-layer, which was built on a Dennis Lancet coach chassis and carried a total of 1½ miles of hose, which could be paid out, one or two lines at a time, at a rate of 15 m.p.h.

Speed was essential, too, for another type of appliance which was becoming increasingly common during the 1930s, the airfield crash tender. The outspoken editor of *The Aeroplane,* Charles Grey Grey, was appealing to designers of the period to develop 'planes that land slowly and do not burn up', but landing speeds – and crashes – were still on the increase. The call was therefore for an appliance which could be on the job within seconds, and here the newly announced Ford V-8 truck was widely used. Henry Ford had pulled off something of a minor miracle in 1932 by adapting the normally expensive V-8 power unit for mass-production cars, and when fitted to the Ford truck range, the V-8 gave a dramatic increase in power and performance: the aviation contractors Airwork were among those who ordered a Ford V-8 for airfield duties.

Other chassis were used as a basis for such crash tenders – Merryweather built this type of appliance under the name Warspite, with a turbine pump for both foam and water on Morris-Commercial and Commer chassis as well as Ford – but nearly all were of the old open Braidwood type. A spectacular exception was the batch of six-wheeled Crossley fire-tenders supplied to the Royal Air Force in 1935. Their tear-drop-shaped bodywork contrasted curiously with their square-rigged bonnet and radiator – 'How the sloping stern, which is not a true streamline, is going to help while cantering across an aerodrome, is not explained,' carped *The Aeroplane* – and the roof of the cab was glazed to give the driver overhead vision. Grüss airsprings were fitted to the suspension to give an easier ride over the irregular surface of a grass airfield.

The Crossleys were to give a display of their prowess at the 1935 Hendon Air Display, when a crazy flying exhibition went sadly wrong:

> Cutting his motor at about forty feet in full view of everybody, Mr Heu-Williams, as the unduly dumb aeronaut, sank unconcernedly towards the ground as if to bounce the machine on its very strong underwork and carry on. We rather expected him to ease the motor fully, but this he did not do, and the machine hit the ground fair and square and disintegrated, without damaging the occupant, who stepped out after a moment . . . This incident gave a wonderful opportunity for the organisers to show the preparedness of the arrangements. Within five seconds of the accident another ham-handed pupil in a similar gamboge-coloured Avro Tutor was already off the ground and carrying on the good work of following the instructor. The new 60 m.p.h. Crossley streamlined fire-tenders appeared on the instant and covered the airframe and motor of the stricken Tutor with fire-foam.

However, the six-wheeled Crossleys subsequently supplied to the Air Force had open and far from streamlined bodywork.

Smaller engines such as the Ford V-8 were also much in demand by private brigades and the less affluent municipalities – in fact, the earlier Ford Model A with Simonis firefighting equipment had also been a popular basis for fire engines from 1931, when one of the first examples, liberally bedizened with chemical extinguishers had been supplied to Bertram Mills Circus. Some Continental makers built engines which could be used for street-watering as well as firefighting like the Ford V-8 converted by Interprindere de Maşini şi Automobile SA, of Arad, Roumania, which was sold to Istanbul in August 1938.

Intermittent attempts to build even smaller appliances on tiny chassis like the Morris Eight and the Gwynne Eight (whose makers were also famous for their turbine pumps) were doomed to failure after only a handful of examples had been built. In contrast, some of the medium-sized engines supplied during the 1930s were still in service in the 1970s, especially in Australasia and America, though not all these old timers were regarded with sentiment by their users. Take the Blayney, New South Wales, volunteer brigade, who in the late 1960s were clamouring for a more modern appliance to replace their 40-year-old Garford engine. It had, they said, a 20 m.p.h. flat-out maximum speed, and was lucky to reach 4 m.p.h. on hills; its siren was used to scare off children who persisted in overtaking the struggling machine on their bicycles.

If the interwar period saw a rapid advance in bodywork design, the progress in the design of firefighting ladders was also impressive. Until then, perhaps the greatest design advance had been the replacement of the hinged 'fly-ladder' devised in the 1830s by Abraham Wivell of London by the telescoping extending ladder in the 1880s. In 1892 the German firm of Magirus built a turntable ladder unit on a horse-drawn trailer, and in 1906 they built the first motorized turntable ladder. Their lead was followed in 1908 by Merryweather who built the first British engine to use its road motor also to provide power to elevate and extend the ladder; this was sold to Shanghai. Chief Officer Lee Tuppen, of the Rangoon Brigade (who was a former Merryweather engineer) suggested to his old company in 1924 that they should build him a motor turntable unit in which the ladder was elevated hydraulically. As delivered, the engine featured a three-throw hydraulic pump at the rear of the chassis, driven by the power unit and developing a pressure of

(Below) A glistening 1930s model on show in Canada

450 p.s.i., which was adequate to raise and extend the 93-ft-long ladder.

These turntable ladders could be used to fight fire in tall buildings, water being pumped to a 'monitor' nozzle at the head of the ladder, initially by another pump, later by a pump built into the turntable vehicle itself. Demand for such units naturally grew as buildings became higher, and a major development came in 1931 when Peter Pirsch built the world's first all-powered aerial ladder truck, for the city of Spokane, Washington. The ladder turntable was directly over the swan-neck coupling which linked the trailer to its towing vehicle, and, like so many of these extra-long American units, the trailer had steerable rear wheels and its own steersman.

The same year, the German company of Magirus, which had been active in the fire appliance field since 1864, produced a turntable ladder which had an extended length of 150 feet. Built entirely of steel, the ladder was shown at the International Fire Apparatus Exhibition in Paris, 'where it won unbounded approval'. In April 1932 the Manchester fire equipment company of John Morris supplied a 100-foot Magirus all-steel water tower and turntable ladder to the London Fire Brigade; this was claimed to be the first mechanical ladder of this type to be used in Britain.

Leyland signed an agreement in 1935 with the German Metz concern, which could trace its origins back to 1842, and which had built its first turntable ladder in 1912. Metz patented a hydraulic turntable ladder in 1935, and it was this which Leyland had contracted to use on appliances sold in the British Commonwealth. An ingenious feature of the Leyland-Metz escapes was an automatic stabilizing device to prevent the ladder from being over-extended and over-balancing the appliance. Such a fitment was very necessary on the taller Metz units, such as the 45-metre unit supplied to Hull in 1936, which was the longest ladder which had been put into service with a British brigade.

But it was New York, where there seemed no limit to the height to which the skyscrapers would soar, which had the greatest need of special firefighting appliances. In 1930, American LaFrance supplied the city with a water tower 65 feet high which could deliver the astounding total of 8500 gallons per minute through four nozzles; in 1933 came a pair of pumpers purpose-built for fighting fire in the newly completed Empire State building. Featuring four-stage pumps which developed a discharge pressure of 600 p.s.i., these appliances could deliver water at a pressure of 250 p.s.i. at the very top of the Empire State, 1250 feet above street level. But they were never delivered: they became a pawn of New York's city politics, the sale was stopped, and the two super-orphans were back with their makers, who fortunately managed to find homes for them in Ithaca, New York, and Youngstown, Ohio. Incidentally, American LaFrance were now building their own V-12 engines, which proved versatile (and powerful) enough to be fitted in vehicles as diverse as Greyhound buses and US Army tanks as well as boosting the pumping capacity of American LaFrance fire engines to 1500 gallons per minute.

(Far left) Still in the traditional mould: a Leyland of 1933 vintage

(Left) Merryweather also built their Warspite engine on Bedford chassis: this 1935 example was used by the Bristol Aeroplane Company

In 1942, American LaFrance built a 125-foot aerial ladder to go into service at Boston Massachusetts; the following year it was working at a fire when suddenly a wall collapsed and completely flattened the truck. But the ladder stood 'fully extended and safe' despite the accident.

In the same year, Japan delivered her lightning attack on the undefended Pearl Harbor. Not long afterwards, America joined the rest of the world at war. And with war, came a new horror: *Blitzkrieg*.

FIRE ENGINES GO TO WAR

Despite the technological advances of the 1930s, as the Second World War loomed nearer, the firefighting services of Britain were found to be in a sorry state. As early as 1932, Marshal of the Air Force Sir Geoffrey Salmond had forecast: 'In the event of hostile action from the skies, the task of limiting the damage will devolve on the Fire Service'.

But the country's fire defences were a piecemeal agglomeration consisting of many hundreds of brigades numbering only 16,564 men, of whom one quarter were full-time firemen, just over 2000 policemen 'available when necessary for firefighting duties', 4746 'retained' men, and the remainder volunteers. Equipment ranged from the antique to the ultra-modern, fire stations from the primitive to the palatial. One brigade, it was said, had refused to attend a serious fire in 1930 because its members were ashamed of the inadequacy of their equipment; other penurious brigades had to advertise for 'reach-me-downs' in the firefighting press. One Sussex group, unable to find proper firemen's tunics, was reduced to wearing secondhand conductors' uniforms from the local bus service.

So in 1938 the Fire Brigades Bill was passed by Parliament, compelling all local authorities above the parish level to provide a fire brigade, either separately, or jointly with neighbouring authorities. The 'parish pump' ceased to exist, having outlived the last insurance brigade by a decade.

Spurred by the Munich Crisis, the Home Office frantically commissioned a million pounds worth of new equipment, though they should have ordered as much again: and an Auxiliary Fire Service was formed, which by the outbreak of hostilities could number some 100,000 men. As the 'Phoney War' period dragged on, however, and the enemy bombers did not come, public criticism of the AFS became bitter, and many of the auxiliaries left the service to join the armed forces.

Then, in the summer of 1940 came the Blitz: Croydon Airport was bombed, incendiaries fell on Woolwich and Eltham, the West India Docks were set on fire. All these preliminary raids were successfully brought under control, but the folly of allowing the Auxiliary Service to run down to half-strength was shown, and at the height of the Blitz men and women had to be conscripted into the AFS.

The German raids on London started in earnest with the burning of Shell Haven oil refinery on the night of 5 September; after many difficulties (mostly caused by the bureaucracy engendered by the 1938 Act) the fire was brought under control, only to be rekindled by further enemy attacks the following afternoon. That evening, 600 aircraft attacked Dockland, and the London Fire Brigade had to deal with over 1,000 fires.

For the next 57 consecutive nights, enemy bombers blitzed London; the Brigade managed to deal with all the resulting fires. In November the raids ceased while the Luftwaffe turned its attention on other cities – Coventry, Birmingham, Liverpool, Sheffield among them – but in December London was once again under attack.

It was said that the Germans had been taking careful note of the tidal ebbs and flows in the River Thames; certainly the river was abnormally low on the night of December 29 when the enemy bombers dropped a claimed 100,000 incendiary bombs on the City of London. In fact, it was impossible for the Brigade to use river water because access was too difficult; and their problems were confounded when high-explosive bombs shattered a dozen of the City's biggest water mains. It became necessary to use long hose lines to bring water from a distance, and one fire a quarter of a mile square in the Minories became uncontrollable through lack of water. That night 1500 fires were fought, and the central control of the firefighting services had to be moved twice as fire stations were surrounded by fire and had to be abandoned. When firemen fighting blazes in one city street were hemmed in by fire, they left their appliances and escaped through the tunnels of the underground railway.

That night, London had 2000 pumping appliances in action, augmented by another 300 from outside the city, plus over a hundred water-carrying tenders and nine fireboats. By 4.30 am the impossible was achieved: all the fires were under control, though at the cost of fourteen firemen dead and 300 injured.

A curious assortment of vehicles was in use at that period: some auxiliary brigades created their own fire engines from cut-down cars or trucks, with the addition of a trailer pump. There were also many properly-designed auxiliary towing vehicles, but to supplement these, at the outbreak of war, many of London's taxis were commandeered as towing vehicles. The trailer pumps enabled the firefighting services to bring far more extinguishing power to bear than if they had just relied on conventional fire engines, and were produced in many forms, ranging from heavyweight giants capable of delivering 700 gallons per minute to 140 g.p.m. lightweights. One of the most popular pumps of the period was the Coventry-Climax, which used the engine which this company had designed for the Swift Cadet light car: when Swift went out of business in the early 1930s, Coventry-Climax turned the disaster to good account by adapting the now surplus-to-requirements power unit for use in their new fire pump, and succeeded so well that some 25,000 of these were eventually supplied. Indeed, some of those little wartime 'wheelbarrow pumps' were still in use in 1976: 'Their lightness and portability makes them the best thing we have for basement fires', commented a London fireman.

(Previous page, inset) Bomb damage at John Lewis's store in Oxford Street, London

(Previous page) London's dockland in flames after the first mass air-raid, September 7, 1940

(Opposite) Scenes from the London Blitz: 'the heroism of London's firemen . . . is clearly illustrated in this striking set of pictures' said the original caption to these photographs

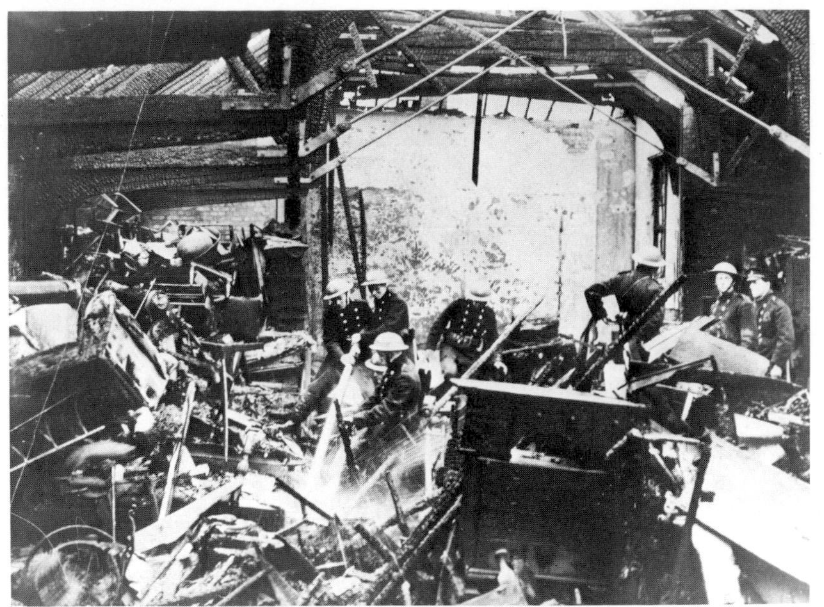

A variation on the trailer pump was the Ford-Sulzer unit, a cradle holding a Ford V-8 engine and pump which could be carried in the back of an auxiliary vehicle, or carried closer to the fire. During the Blitz, Ford-Sulzer pumps were installed on the bridges of London to suck up water from the Thames no matter how low the tide.

During the last great raid on London in 1941, on 10 May, when 500 fires were started before midnight, nine miles of hose had to be laid to combat a huge blaze at the Elephant and Castle, where water was drawn from the local swimming baths, from the Surrey canal and from the Thames.

It was twelve days before the last appliance was 'stood down' after this attack.

Though the firemen had achieved miracles in dealing with the German Blitzkrieg tactics, the organization of the various brigades round the country was still far from perfect. There was a general lack of co-ordination – hardly surprising when you consider that there were still 1400 different brigades nationwide – and poor communications often meant an unnecessary increase in damage and casualties when one brigade had to be called to the aid of another. So, with great speed, the National Fire Service was formed on 18 August, 1941 – excessive speed, in one respect, for it was discovered three years later that the NFS had been constituted so rapidly that the necessary Parliamentary procedure had not been followed, and that, therefore it was illegal – to concentrate the nation's firefighting strength wherever fires might occur. To this end, the country was divided into 11 regions, made up of 33 fire areas. All the brigades in a fire area were grouped into one force: in the case of major fires, the Chief Regional Fire Officers and the Chief Commander in London were empowered to take control.

The whole reorganization took less than three months, and generally with remarkable ease; but the Luftwaffe had done its worst, and in fact the National Fire Service was not called upon to show its full strength in emergency. During the flying-bomb attacks of 1944–5, NFS men attended every Doodlebug incident in London, carrying out salvage and repair work and evacuating the homeless, as well as dealing with firefighting and rescue work.

During the war the London Brigade alone dealt with over 50,000 incidents; at its peak the London Region Fire Service had 42,000 men and 10,000 vehicles on its strength.

In 1948 the National Fire Service vanished by Act of Parliament, and the responsibility for providing fire defence fell largely upon the counties and county boroughs of Britain, who had to carry 75 per cent of the cost – estimated at £13,750,000.

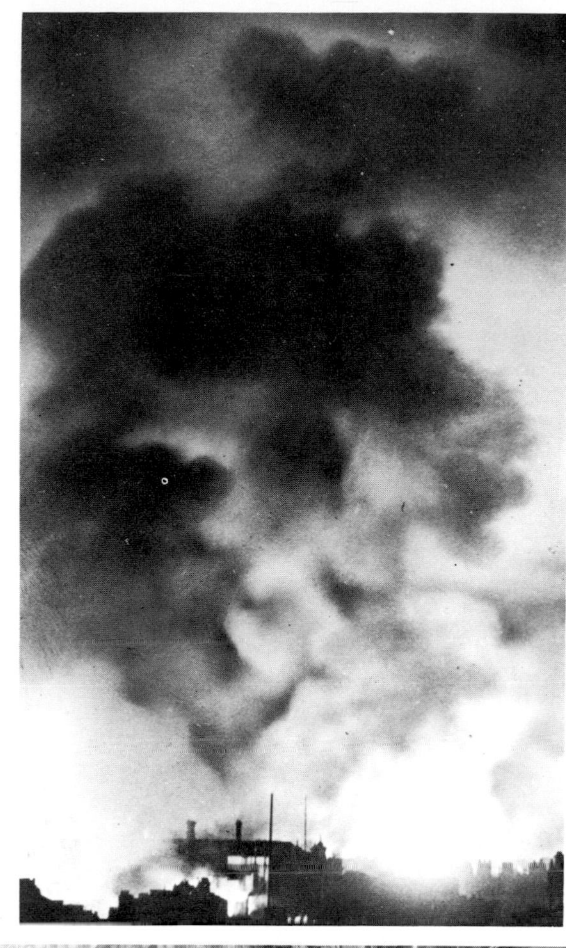

During the Blitz, London's firemen were really fighting in the 'front line' against the Luftwaffe – and their bravery often cost them their lives.

Though America had escaped the bombing raids which had caused so much damage in Britain (and, indeed, in Germany), the citizens of New York had experienced something of the feeling caused by an aerial bombardment in the summer of 1945, when a Mitchell bomber, flying at 300 m.p.h. in poor visibility, crashed into the 78th and 79th floors of the Empire State Building. The wings were torn off; one of the engines smashed down on to the roof of a neighbouring building and started a fire. The other engine and the fuselage punched through the north wall of the Empire State Building releasing a stream of blazing petrol which cascaded down the stairs to the 75th floor. Staff firemen managed to keep the flames in check until the New York Fire Department arrived, breathless no doubt after having to carry their equipment from the 67th floor because the lifts were out of action. In just over an hour, the world's highest fire was under control.

THE MODERN ARMOURY AGAINST FIRE

This Faun foam cannon with eight-wheel-drive operates at Frankfurt Airport, and can cope with burning Jumbo Jets

One of the most remarkable post-war developments in firefighting technology has been the development of the 'Super-Pumper' for dealing with large fires in urban centres. Like many modern ideas, it has its echoes in history: in the late 1930s, American LaFrance built a special Metropolitan Duplex outfit for Los Angeles, which had two V-12 engines, for motive power and for pumping and twin two-stage pumps with interconnected inlet and discharge passages, which supplied a 'companion truck' which had more than 20 outlet points. It was an interesting attempt to increase the volume of water which could be directed on to a fire: but high-pressure operation was not really practicable at that time, for hoses were not capable of taking high pressures for long periods. Though woven hose had supplanted the old, temperamental riveted leather type in the mid-nineteenth century, until the Second World War, little had been done save to make hoses more watertight by lining the woven flax or cotton jacket with rubber. Then, synthetic materials such as nylon gave the ability to make hoses that would not rot and that would withstand greater pressures, leading to the strong, all-synthetic hoses of the 1960s and 1970s.

One of the leading bodies in the development of the high-pressure hose was the United States Navy: and it so happened that the British Navy was simultaneously developing a lightweight, high-speed diesel engine of unusual configuration. This was the eighteen-cylinder Napier Deltic, which had its cylinders arranged in a triangle, with two pistons per cylinder, and could produce 2400 b.h.p. at 1800 r.p.m. Mack Trucks had the idea of uniting these two separate streams of development, and the result was the Super Pumper, Super Tender and Satellite Tender complex supplied to the New York Fire Department in 1965. Heart of the complex was the Deltic-powered Super Pumper unit, housed in a semi-trailer drawn by a Mack

(Left) Firemen at a blaze in the City of London

(Below) U.S. Navy firefighters deal with a crashed fighter plane

tractor vehicle; the Deltic drove a six-stage DeLaval centrifugal pump, which could either deliver 8800 gallons per minute at 350 p.s.i. or 4400 gallons per minute at 700 p.s.i. Taking water from the most inexhaustible supply available, the Super Pumper fed it through either four or eight 4.5in high pressure hoses to the Super Tender (another artic combination, but this time armed with a 10,000 g.p.m. water cannon, and carrying tools, equipment and hose within its trailer body) and the Satellite Tenders, four wheelers equipped with 4000 g.p.m. water cannon and 2000 foot of hose. The combined force of the Super Pumper and its satellites could hurl up to 37 tons of water on to a fire every minute.

Even so, buildings in city centres, where land costs were high, were soaring upwards far beyond the reach of even the most powerful pumpers. It was a problem dramatically brought before the public in the film *The Towering Inferno*, but in reality the solutions had been carefully thought out by civic

Sao Paulo disaster – The 'Towering Inferno' became tragic reality on February 1, 1974, when one of the newest office blocks in Sao Paulo, Brazil, the 25-storey Joelma Building, went up in flames. The city's fire brigade found that its ladders were too short and its hoses not powerful enough to reach the upper floors. But the real cause of the tragedy, in which 227 people died, was the use of inflammable plastics for interior paintwork and panelling and the lack of adequate escape stairways. Once the fire had taken hold, the only escape for those on the upper floors was to the roof, where there was a helicopter pad. However, it was two hours before the flames had abated sufficiently for a helicopter to land, and by that time only 80 people had survived. Some of those trapped by the fire were rescued by lines fired from neighbouring

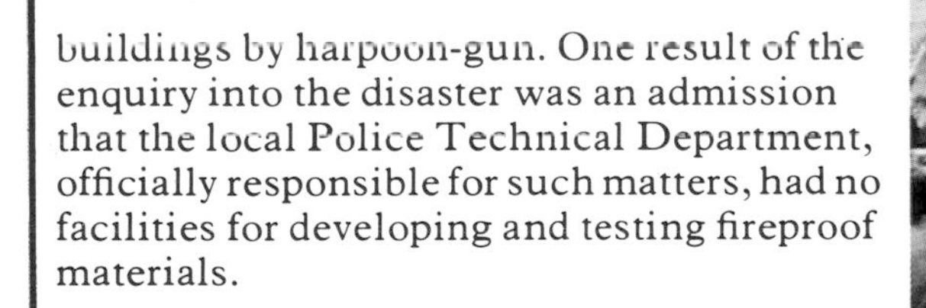

buildings by harpoon-gun. One result of the enquiry into the disaster was an admission that the local Police Technical Department, officially responsible for such matters, had no facilities for developing and testing fireproof materials.

authorities. Commented a London Fire Brigade spokesman: 'Under the London building acts, all the appropriate firefighting systems have to be built into the structure of high-rise buildings. These can take a number of forms – dry risers, wet risers, and other appropriate means of prevention. Everything has to be part of the building'

Another powerful modern weapon in the armoury against fire is the Snorkel, a high level rescue and monitor platform of far greater versatility than the turntable ladder. Irreverently known as the 'cherry picker', the Snorkel has an articulated, hydraulically operated arm carrying a four or five man working cage, which can be elevated to a maximum height of 85 feet. The men in the cage can control its movements hydraulically with great precision, and though the turntable ladder is still supreme where sheer height is all-important, for many jobs the Snorkel has the edge: a Snorkel cage has a safe working load of 800 lbs, while only one man at a time can stand at the top of a turntable ladder.

Midway between the Snorkel and the turntable ladder comes the Mack Aerialscope, with a working cage at the top of a telescoping boom which can be moved through an arc from 10 feet below ground level to 65 feet above ground level, or rotated through a complete circle.

There are, of course, some modern developments which, though not adopted for general use at present, promise great things for the future. In 1966 a hovercraft fire engine made its debut at an exhibition given by the Royal Air Force near Portsmouth. The hovercraft carried a lightweight pump, two lengths of lightweight suction hose, six lengths of discharge hose, foam-making equipment, a 200 lb dry powder unit and a 200 gallon water tank. A couple of years later, it was reported that the London Fire Brigade was considering putting hover fireboats into use on the Thames, and design studies were drawn up by Merryweather.

(Opposite page) French sapeurs-pompiers in action

Fire at sea is, indeed, a specialized branch of the firefighting art. The fact that a ship is surrounded by water can prove a two-edged sword, as New York firemen found in 1942 when the liner *Normandie* caught fire while docked in New York harbour. They pumped so much water into the ship to extinguish the fire that the *Normandie* sank under the weight.

Another famous ship gutted by fire was the original *Queen Elizabeth*, which burned in Hong Kong harbour while being refitted as a floating university.

On the high seas, ships must look to themselves for salvation, though not always with success, as the case of *l'Atlantique*, burned out beyond recovery in mid-Atlantic in 1933, proves.

Nowadays, the increasing size of oil tankers gives frequent cause for concern: putting out a blazing 500,000 tons tanker could be a daunting prospect.

(Above) Fireboats in action at a dockland fire

(Opposite page) A training session for London firemen

Another form of firefighting unit which is still undergoing development is a jet engine adapted to produce an inert gas to quench flames. Tests have shown that such an engine can generate enough gas to fill a large building within a quarter of an hour. Jet engines can also be used to generate high expansion foam, which is expected to have wide firefighting applications in the future.

Some modern methods of firefighting already use techniques that would have been regarded as impossible only a few years ago. The flying fire engine, presaged by the San Diego experiments of 1918, is now a standard weapon in fighting forest fires, especially in Australia and North America. 'Bombing' forest fires with water was first developed in Canada shortly after the last war, and has been developed to the extent that aircraft can now drop 6000 gallons of water at once. Helicopters are used in the United States as flying hose layers, and aircraft have also been used to dump fire retarding materials ranging from chemicals to mud on forest fires.

One of the most demanding tasks faced by modern firefighting vehicles is that of dealing with crashed aircraft: it has been said that an aircraft crash has to be dealt with within sixty seconds from impact – and this includes getting perhaps 200 to 300 passengers safely out of the fuselage. The design of aircraft crash tenders is therefore highly specialized, and these vehicles normally have four-wheel – or sometimes even six-wheel – drive for maximum traction across the grass of the airfield. Such tenders carry foam cannon which can discharge up to 13,500 gallons per minute, a facility which is sometimes used to lay a foam carpet on which the aeroplane can land in the event of an undercarriage failure without fear of sparks igniting the aircraft's fuel supply. Perhaps the ultimate aircraft crash tender is the German Faun with a 1000 h.p. Daimler-Benz V-10 diesel engine and eight-wheel drive intended to cope with Jumbo Jets carrying up to 500 passengers.

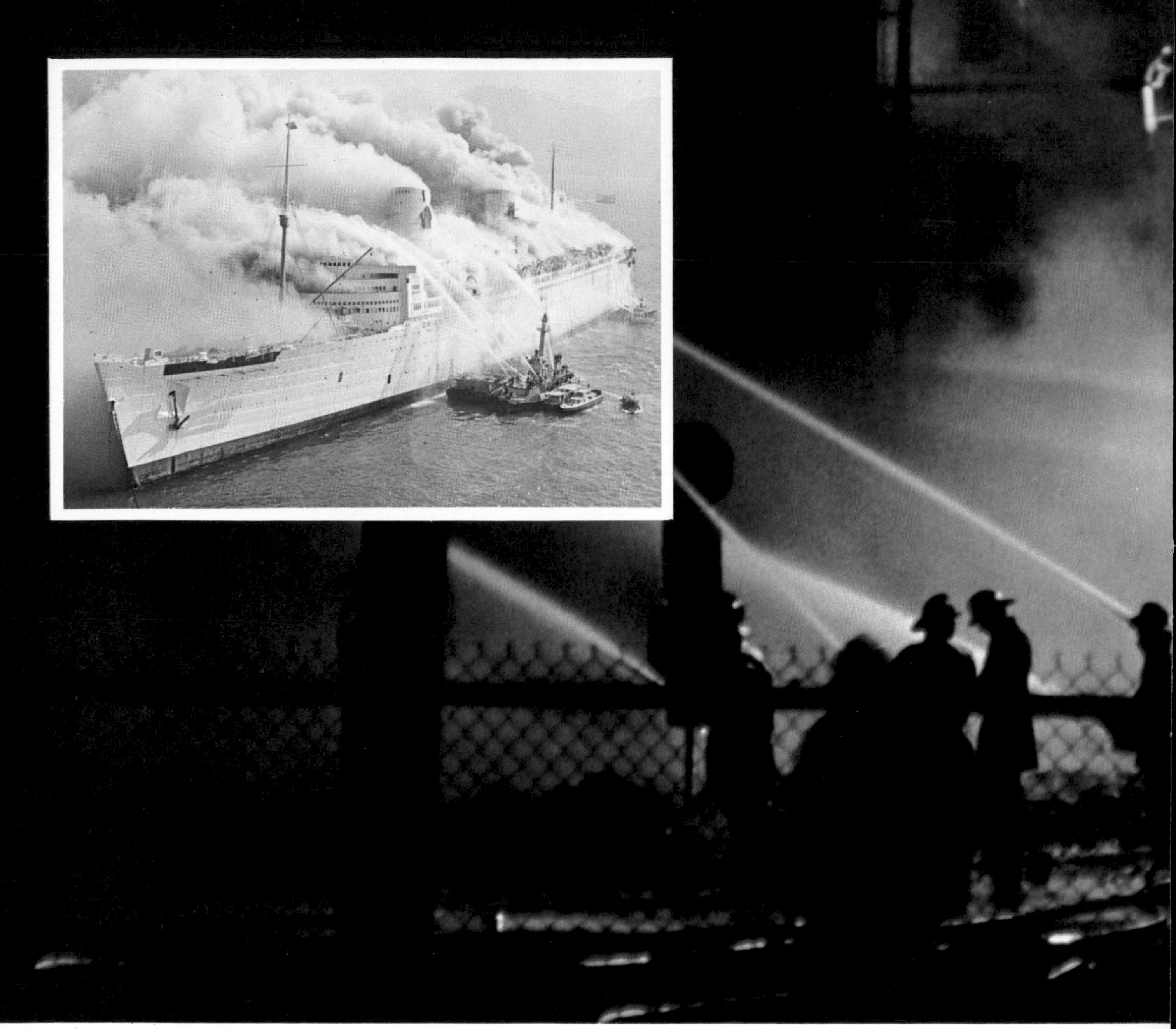

The liner Queen Elizabeth **on fire in Hong Kong harbour**

(Above) **New York firemen tackle a night-time blaze**

But these are the glamorous specialist vehicles: what is the typical all-purpose fire appliance of today like?

It is, in most cases, powered by a diesel engine, since diesel fuel poses less of a fire risk than petrol. Furthermore, diesels are more tolerant of poor atmospheric conditions and can be fitted with special breather packs for smoke laden air. However, in essence, the modern fire engine is a logical development of its forebears, though it may be larger and more powerful.

Fire engine makers still offer a bewildering variety of appliances. They have to, because each major fire poses a different problem. Take, for example, the variety of equipment operated by one of the longest-established brigades, that of Edinburgh in Scotland, which includes hose layers, salvage tenders, emergency tenders and high expansion foam trailers alongside pumping engines of conventional layout and Snorkel units. The brigade even boasts a fully equipped mobile fire control unit which is, in fact, an office on wheels designed to superintend firefighting operations on the spot, working in conjunction with the central control at the brigade headquarters. The facilities provided by this vehicle include radio communications between brigade headquarters and the firemen's walkie-talkie radios, a tape recorder to log information about the progress of the unit's fight against the outbreak, and control for the breathing apparatus worn by the firemen. Carrying its own generator, this is a completely self-contained unit.

Among chassis makers, competition to supply the fire engine builders is keen: many truck manufacturers offer a specially adapted version of their standard product for this purpose, though a few manufacturers, like Peter Pirsch in America, still prefer to construct their own chassis (though Pirsch appliances can also be ordered on a standard Ford chassis). Some builders will even provide a chassis to bespoke order, like Ford, who

A typical modern American fire engine – a 1972 1000 gpm Imperial. But whatever has become of the traditional 'fire-engine red'?

A modern Ward-La France ladder truck

(Right) Today's firefighters may have to cope with chemical fires like the one which devastated the Nypro works at Flixborough, Lincolnshire in 1974 – Britain's worst industrial disaster

(Above) Using a derelict DC-3, airport firemen demonstrate the power of chemical foam in bringing burning aircraft under control

(Overleaf) Firemen tackle a blaze in the Salmon National Forest, Idaho: fires like this are all too often the result of human carelessness

claimed that consultations with Fire Authorities, the Home Office and leading fire appliance manufacturers had enabled them to develop a 'chassis package which could satisfy all known operating conditions, and to aid the fitment of bodywork and lockers by complying with certain measurements specified by the fire engineers'.

So the trend seems to be that new forms of firefighting machinery will continue to be developed alongside refined pumpers and turntable ladders. Firefighting has become more of a science, less of a battle with the unknown, though the fireman still has to rely on his traditional skill and courage as much as he did in the days of Braidwood and Massey Shaw.

In fact, one of the most colourful descriptions of a fireman's duty still has many points of truth today, though it was addressed to New York recruits in the 1880s:

'Gentlemen: You have been chosen from among 800 applicants, and I expect you all to be sober, industrious, and honest; and I also expect that you will obey all orders with alacrity and willingness. Avoid all discussions with your fellow labourers, and do all your work without grumbling. Politics and religion are subjects which I positively forbid being discussed. Ignore them absolutely. Should you become involved in a misunderstanding with a fellow member of your department, come to me and I will arbitrate your difference at once. Be sober, for if you are drunk your brains are out, and you are no longer fit for duty. Drunkenness will certainly not be tolerated. In your whole deportment, show yourselves to be gentlemen. I consider you such, and there is no reason why you should not act as gentlemen at all times. Profanity is uncalled for. It is a vile habit, and one which I have always got along without. I never practice it, and hope that you will follow my example. Be polite: and now report at your posts.'

30

INDEX

ACKNOWLEDGMENTS

The publishers would like to thank the following organizations and individuals for their kind permission to reproduce the photographs in this book:

Associated Press 110 below right, 113 below, 114, 119 left and right, 121; Birmingham Fire Brigade 91 below; B. L. Blunt 91 above; David Burgess Wise 68, 73 above, centre and below, 75, 76, 76 inset, 77 left and right, 79, 85 above, 92, 93 above and below, 95 centre; Camden N.J. Fire Dept. 57; Colourviews 12 below, 17, 46, 53, 67 above and centre, 69 above and below, 70, 71, 81, 96–97, 101 above; Daily Telegraph Colour Library 38–39; Greater London Fire Brigade 12 above, 31; Guildhall Library 6–7, 15, 22, 32; Illustrated London News 54–55, 58, 59; Keystone Press Agency 108–109, 110 top, 110 above and below left, 112 below, 113 above, 125; Library of Congress 42, 44–45, 50, 56, 60–61, 63 centre and below, 64–65, 83, 90; London Fire Brigade title, 65, 67 below, 78, 80, 84, 117 above; London News Agency 94; London Salvage Corporation 38 inset; Roger Mardon 102 above (inset), 106, 106–107, 123 below, 124–125; Melbourne Fire Brigade 97 inset; Merryweather Museum 82 below; National Motor Museum 93 centre; Photri ends, contents, 62–63, 86–87, 88–89, 117 below, 122–123, 126–127; Popperfoto 108, 110 centre, 112 above, 122; Rex Features (Champlong Arepi) half-title, 114–115, 118; Royal Exchange Assurance 14; The Director, The Science Museum, London 8–9, 10, 13, 18–19, 20, 37 above and below, 48–49; Smithsonian Institution 21, 23, 24–25, 30, 35 above and below, 72; Syndication International 124 below; Thompson 16, 33, 34, 41, 82 above, 85 below, 95 above, 96 inset, 98 inset, 98–99, 100–101, 101 centre and below, 102 centre (inset), 102–103, 104, 107, 116, 120; ZEFA (Photri) 26, 27 centre and below.